A/V TROUBLESHOOTER

Audio-Visual Equipment Operation, Maintenance, and Repair

Lesley Kamenshine

A SPECTRUM BOOK

Prentice-Hall, Inc., Englewood Cliffs, New Jersey 07632

ent operation,

—Maintenance and repair.
ubleshooter.
TS2301.A7K36   1985   621.38'044   84-16096
ISBN 0-13-054529-5 (hard)
ISBN 0-13-054511-2 (pbk.)

© 1985 by Leslie Kamenshine.
All rights reserved. No part of this book may be reproduced in any form or by any means without permission in writing from the publisher. A Spectrum Book. Printed in the United States of America.

ISBN 0-13-054529-5

ISBN 0-13-054511-2 {PBK.}

Editorial/production supervision: Marlys Lehmann
Cover design: Hal Siegel
Manufacturing buyer: Pat Mahoney and Anne Armeny

This book is available at a special discount when ordered in bulk quantities. Contact Prentice-Hall, Inc., General Publishing Division, Special Sales, Englewood Cliffs, N.J. 07632.

Prentice-Hall International, Inc., *London*
Prentice-Hall of Australia Pty. Limited, *Sydney*
Prentice-Hall Canada Inc., *Toronto*
Prentice-Hall of India Private Limited, *New Delhi*
Prentice-Hall of Japan, Inc., *Tokyo*
Prentice-Hall of Southeast Asia Pte. Ltd., *Singapore*
Whitehall Books Limited, *Wellington, New Zealand*
Editora Prentice-Hall do Brasil Ltda., *Rio de Janeiro*
Prentice-Hall Hispanoamericana, S.A., *Mexico*

To Bob, Sally, and Wendy—
the special people in my life

Contents

Preface .... *vii*
Acknowledgments .... *ix*

1 General Tips 1
2 Lamps 10
3 Overhead Projectors 13
4 Opaque Projectors 19
5 Dry-Mount Press 22
6 Laminators 27
7 Thermal Transparency Makers 41
8 Cameras 45
9 Visualmakers 78
10 Copy Stands 84
11 Slide Projectors 86
12 Dissolve Units 100
13 Filmstrip Projectors 113
14 Movie Cameras 124
15 Super 8 Loop Projectors 135
16 Super 8 Movie Projectors 138
17 16mm Movie Projectors 144
18 Tape Recorders 166
19 Phonographs 186
20 Microphones 191
21 Headsets, Jackboxes, Listening Stations 197
22 Connectors, Jacks, and Plugs . 202
23 Public Address Systems 209
24 Television 214
25 Videotape 218

Bibliography *253*
Index *257*

Preface

Welcome, particularly if you haven't used much audio/visual equipment before, or if you don't like machines. Yes, you can ... touch the machines, use them successfully, even enjoy them and deal with foul-ups. I am talking about any of the types of equipment discussed in this book.

While your equipment may or may not be the latest model, my basic premise is that you have misplaced your operating manual, and it is the weekend. You can adapt the information provided in this book to the make and model of your equipment. The book is written in a simple outline form for quick reference. Almost every chapter has four parts: Operating Tips, Problems, InDepth, and Glossary.

The beginning of each chapter explains the purpose for which the equipment is used and its basic features.

Following are operating tips, which include reminders, not only for the equipment but for the film, tape, or slides you are using—the software. Did you know, for example, that unused film is good for six months after the expiration date marked on the box? Are you aware that the same videotape can record both black and white and color, that it's the camera that makes the difference? Or that most equipment must warm up for one hour after being in extreme cold, or it may not function properly?

The heart of each chapter is the troubleshooter's guide, the problems section. The most common, and sometimes not so common, problems I and others in the field have experienced are listed along with their causes and solutions. You may need to know how to unjam a laminating machine, deal with an unsynchronized slide–sound program, or fix garbled sound on a movie projector.

The InDepth section shows the parts of the machine and explains operating procedures, maintenance, and some little-known applications to help you use the machines and software creatively. Some suggestions appear in illustrations, and some are in the text.

Almost every chapter has a glossary. I have listed terms important to each type of equipment so that you can be more knowledgeable when you talk with the mechanic who makes the big repairs, or so that you can understand those who are showing you how much they already know.

There are two special chapters, General Tips and Lamps. General Tips is the first chapter of the book. It is well worth your time and will help you to avoid future difficulties. What kind of instant-help kit should you have on hand? How can you prevent harmful heat buildup on your machine? How do you clean batteries and terminals, and how do you handle rechargeable batteries? How do you clean screens, and how far away should the equipment be from the screen? How can you tell the model number of your machine?

Lamps is a short chapter. Projector lamps have special codes. Do you know how to order projector lamps? How you prevent them from burning? The chapter provides a troubleshooter's guide.

Operating equipment is a serious matter. The equipment is expensive, and your time is valuable. But with a twinkle in your eye and *A/V Troubleshooter,* your presentation will be a success.

Acknowledgments

Over the years many people have helped me. Technicians have checked facts; amateurs have read over the material to see if it met their needs.

Three people first spurred me on to write this manuscript and provided invaluable advice: Willodene Scott, director of libraries, and Earl Stiles, formerly supervisor of the audio/visual department, at the Metro Nashville Public Schools, and my friend Ben Zucker. Also Duane Muir, professor and director of the audio/visual department at Nashville State Technical Institute, and Steve Adams, technician extraordinaire at Consolidated Media Services in Nashville, read and reread the manuscript. Emily Jorgensen, Peggy Montgomery, Barbara Swift, and Vicki Swift were continuing sources of creativity and especially friendship. Thank you to Ann Bishop, who so accurately, promptly, and cheerfully typed and updated what you are reading.

My family—Bob, Sally, Wendy, Lindsay, and Ruffles—were the ultimate sources of comfort, patience, and support. This book is more meaningful because of them.

A special salute goes to editor Marlys Lehmann of Prentice-Hall whose skill, patience, and perseverance guided this book through the difficult production process.

Many thanks also to the people listed below, who contributed their time and comments to this project. *Technical assistance:* Bill Brose, John Chastain, Bob Davis, Stanley Earhart, John Freund, Bill Hammond, Tom Hamner, Harold Martin, Ray Orloff, Joe Snodgrass, Ray Tarpley, Bernadette Zaionchkovsky. Thanks also to Larry Bice, Jay Chesley, Steve Bradley, Mark Maguire. *Amateurs who field-tested chapters:* Members of the Nashville Panel; librarians Alice Ewing, Wilma Tice, and Wanda Ryan; Earl Stiles's summer-session class of fifteen teachers learning about audio/visual equipment; Duane Muir's forty Nashville Tech students who are learning to be technicians.

1 General Tips

CONTENTS

Planning Your Presentation	1	Wall Plugs and Cords	7
Instant Help Kit	1	What You Need to Know About	
Handling the Machines	2	Shelf Life	8
Batteries	3	Storing Equipment	8
Projector-to-Screen Distance	4	How to Determine the Model of	
Screens	5	Your Machine	9
How Much Equipment One Wall		Cleaning and Servicing Equipment	9
Outlet Handles	6	Equipment Repair	9

PLANNING YOUR PRESENTATION

1. Decide which medium would be most effective in conveying your message to the audience. It could be filmstrip, slides, movies, video, or some other medium. Check chapter introductions for equipment usage.

2. Find out whether this equipment is readily available. Check schools, public libraries, university media centers, and local clubs.

3. Reserve the equipment for the time you will need it.

4. Develop your own presentation, or use commercially prepared materials. Material may be copyrighted; unauthorized use or duplication may be contrary to U.S. copyright laws.

5. Become familiar with the operation of the equipment.

6. Practice your presentation beforehand.

7. On the day of the program, allow plenty of time to set up your equipment and your presentation.

8. Be prepared for possible foul-ups. Refer to the rest of this chapter.

INSTANT HELP KIT

Make sure you are well equipped with:

- Spare reels for film.
- Spare reels for tape, an extra reel of tape, or blank cassettes.
- Extra lamp for equipment you will be using, or check out the spare that was included with equipment, since a burnt lamp is often put in place of a spare.

- Heavy-duty extension cords.
- Adaptors for plugging a three-pronged plug into a two-hole wall outlet.
- Masking tape to secure cords to the floor so people will not trip. For heavy traffic, use two-inch gray duct tape, available at hardware stores.
- Extra slide mounts to remount warped slides (slip-in type).
- Miniature screwdrivers—Phillips or flat—available at camera stores, electrical supply stores, hardware stores, and audio outlets.
- Camera lens tissue and camel's-hair brush for cleaning lenses. Do not use eyeglass tissue.
- Canned air for blowing away dust and other particles from hard-to-reach places (available at camera stores).
- Extra batteries—know your size and shape (thin ones, thick ones, round ones?) Where batteries appear the same, know the voltage and type, such as silver, mercury. Use batteries for cameras, light meters, flash units, visualmaker cameras, movie cameras (motor drive, light meter, sound), cassette recorders, video recorders, portable PA system.
- Flash cubes or flipflash, extra film.
- Alcohol and cotton swabs, for cleaning of metallic surfaces. Buy 95 percent isopropyl alcohol, not isopropyl rubbing alcohol.
- Ruler, preferably steel, cork-backed (will not slip), for measuring, tearing, drawing with ink.
- Exacto knife or razor blade. (Holders for blades are available.)
- Tweezers, to remove film from projectors and jammed tape from recorders.
- Blank newsprint, handy as a gluing surface and as protection when using the dry mount press. Check school supply or art supply stores.

HANDLING THE MACHINES

- Don't drop or bang equipment. Avoid shock impact.
- Never force parts or equipment that do not fit easily.
- Before turning equipment OFF, allow the fan to cool the lamp, where there is one. Otherwise, do not move equipment until it has cooled.
- If the machine has a cooling fan but it does not work, do not use the machine.
- Many, but not all, projectors have a heat filter glass to prevent heat from burning the film. The heat filter is usually greenish and is located next to the condenser lenses. If you can tell a heat filter is missing, do not use the projector. Projectors using low-wattage lamps or small study-carrel projectors often don't have heat filters.
- Prevent heat buildup in the maching by allowing adequate air circulation, as follows:

1. Do not block the ventilation slots of the machine, as by placing it on a bed or rug.
2. Do not put equipment in a built-in enclosure, like a shelf, without enough air circulation.
3. Where applicable, remove equipment from its case before using.

- Machines may not operate properly when used outdoors in freezing temperatures.
- Allow forty minutes to one hour for the equipment to adjust to room temperature when coming indoors from freezing temperatures or when coming into air conditioning from the hot outdoors.

- When following directions for use of equipment, if you do not at first succeed, start again from the beginning and retrace your steps, rather than starting at the place where you think you may have made a mistake.
- When focusing equipment, if the focus adjustment has been used fully, try moving the machine closer or farther away from the screen.
- When turning on switches, snap them on. Do not try to sneak them on or turn them gently, as this may cause damage.
- Some equipment needs to have power turned ON, and then to be programmed for a specific function, while other equipment goes from OFF to the desired function. Examine equipment before using it.
- When two halves of a picture are on the screen, framers for filmstrip and movie projectors are used to obtain a complete image. They are not always clearly identified or obvious, so you may have to hunt, since there are many kinds that may operate slightly differently. If you cannot improve the framing of your image, see if the framer was used properly.
- Become familiar with the function, location, and operation of the controls on your equipment before turning lights out and making your presentation.
- Always give the equipment and material a test run before the show to make sure that everything is working at its best.
- Reels—whether videotape, audio, or movie—should not be pinching the tape or film running through them.

BATTERIES

GENERAL INFORMATION

1. Check the strength of the batteries before using the equipment, so that the batteries can be replaced if necessary.
2. Batteries have a shelf life. Replace batteries even if unused over a long period of time. Damage to equipment from battery corrosion is not covered under the equipment warranty.
3. Periodically clean battery terminals with a rough cloth to prevent corrosion and poor contact. Clean contacts on equipment with typewriter or pencil eraser. You may instead lightly spray terminals with a contact cleaner, available at electronics supply stores. When you rub gently with a cloth you will clean terminals and coat them.
4. Remove batteries when they test weak. That is when they can begin to leak.
5. Check batteries periodically to see if they are leaking acid into the battery compartment. If so, discard the batteries and wipe the inside of the machine with a clean rag and alcohol and fine sandpaper (number 320 or 400). Damage to amplifier board cannot be corrected.
6. Remove the batteries when the equipment is being operated on AC power for long periods.
7. Remove the batteries if the machine is to be shipped or if it will not be used for several weeks, such as during holidays or vacations.
8. Recognize plus and minus terminals as follows:
    a. They are usually labeled on the battery.
    b. For AA, AAA, C, and D batteries, the center pole on top is positive and the bottom of the battery is negative.
9. Do not mix old and new batteries because they will only last as long as the weakest one.

4 General Tips

10. Do not mix battery types, as alkaline and carbon zinc.

LEAKAGE

- Most batteries leak slowly. This includes rechargeable, alkaline, mercury and silver batteries. If you do not periodically wipe the equipment contacts and battery terminals, a leakage film between the equipment and batteries will reduce current. Leakage is affected by time and strong temperature swings, as when you take the batteries from a hot car trunk into an air-conditioned house.
- While the leakage of these batteries is not ordinarily very harmful, if the battery is abused or if there is incorrect polarity of one cell in a group of six to eight batteries, the leakage and damage can be severe.
- Regular carbon zinc batteries will not leak from lack of use. However, they will begin to leak profusely if not discarded immediately after they have no more power. This will surely occur where you accidentally leave a device on for a long period without turning it off. A heavy-duty battery is less likely to leak as much as a regular carbon battery.

TEMPERATURE EXTREMES

- Very low temperatures slow up battery energy, but energy will return when the batteries warm up. Warm them in your pocket or in your hands, or place them at room temperature for ten to twenty minutes before using.
- Very high temperatures speed up battery deterioration.

RECHARGEABLE BATTERIES

1. Do not discard rechargeable batteries. Nickel cadmium batteries are rechargeable. Rechargeable batteries are so labeled.

2. Rechargeable batteries have a life of around 500 charges. They range in price from a couple of dollars to $300, depending on equipment needs. Nickel cadmium batteries use up energy very quickly. When they have been discharged, they stop suddenly, rather than fading out gradually. Therefore, keep an extra set on hand.

3. The batteries must be charged in a charger designed for the particular type and voltage.

4. Rechargeable batteries should be recharged immediately after each use. If not charged and allowed to run down, they may never again accept a full charge.

5. Do not use rechargeable batteries if the equipment will only be used occasionally. Such batteries need frequent use.

6. If you know that equipment with a rechargeable battery will not be used for a stretch of time, do as follows:

   a. Remove the battery from the equipment and charge fully.
   b. Store the battery.
   c. When ready to reuse the battery, recharge it fully before inserting it into the equipment.

7. A rechargeable battery will only last about one year without use if appropriate precautions are taken, as explained above. If possible, use the battery for about twenty minutes every five or six months, taking the above precautions.

PROJECTOR-TO-SCREEN DISTANCE

A general idea is provided in Figure 1–1 of the distance necessary between projectors and screen. Assume an average-size classroom and a normal projection lens for that type of projector.

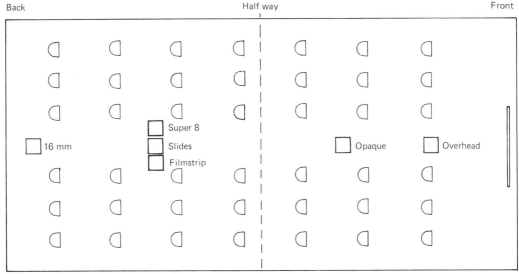

Figure 1-1 *Projector and screen distance. Place projector in center aisle.*

You can make the necessary adjustments once the projector is in the appropriate section of the room:

1. For a smaller picture, bring the projector closer to the screen.

2. For a larger picture, move the projector farther away from the screen.

3. Use the focus adjustment for a clear picture.

SCREENS

SCREEN SURFACES

There are various screen surfaces, each with its own properties. Consider the length and width of your room, the amount of existing light that you will be unable to shut out during projection, and the amount of light from the projector.

FRONT SCREEN PROJECTION

The image is projected onto the front surface of the screen. The projector is located within or behind the audience. This kind of projection is most commonly used for large groups.

When purchasing a screen, inquire about mildew- and flame-resistant fabrics.

Screen-Projector Problem: Keystoning. Keystoning is the distortion of an image on the screen caused by improper relationship of screen and projector. The projector lens and the screen must be perpendicular to each other. If one is tilted too much, keystoning will result. Keystoning can occur with any projection but is particularly prevalent with overheads.

Screen Maintenance. For portable screens, check that metal parts are operating smoothly and that screws on all parts are tightened.

For rips on screen fabric, try iron-on rug-mending tape.

Clean screens with a soft brush or vacuum, or check manufacturer's instructions.

Projection Screen Surfaces

Type	Light Spread or Viewing Angle (Can people sitting at the sides see?)	Picture Brightness and Sharpness
Beaded Beaded surface	Narrow angle; picture best seen when sitting toward front, near projector beam. Not recommended for multi-image presentations.	Bright Needs dark room.
Matt Smooth surface	Greater viewing angle	Not very bright Needs dark room.
Lenticular Ribbed surface	Widest viewing angle	Bright, sharp Room does not need total darkness.
Silver-coated screen (Lenticular)		Extremely bright Room does not need to be very dark.
Translucent surface Primarily for rear projection		Soft, diffused picture

REAR SCREEN PROJECTION

The image is projected onto the back surface of the screen. The screen is between the projector and the audience, and is often built in with the projector. Telex Caramate slide-sound projector and equipment used in study carrels contain good examples of rear screens.

SCREEN MAINTENANCE FOR BUILT-IN REAR SCREEN

To clean the outside surface of a built-in screen, use art gum eraser or mild soap with a moist, soft cloth to wipe the screen. Rinse gently with a clean cloth. Wipe dry. Do not clean the screen from the inside of the machine. This is the repairman's job.

SCREEN MAINTENANCE FOR REAR SCREEN, NOT BUILT ONTO EQUIPMENT

Consult your audio visual dealer or check manufacturer's instructions attached to screen. Some screens cannot be cleaned. Before washing a screen, test a small inconspicuous corner. Some screens have a sprayed coating that can easily be damaged or dissolved. Note that writing like chalk cannot be removed.

HOW MUCH EQUIPMENT ONE WALL OUTLET HANDLES

Electricity is fed from the wall outlet to your equipment. The electrical circuit for one or more outlets can allow only a certain electrical load. Each circuit has a fuse or circuit breaker to cut off electricity in case of an overload.

Similarly, each piece of electrical equipment has a fuse or circuit breaker. In case of an electrical malfunction in the machine, such as a short, the fuse or circuit breaker prevents further electricity from reaching the machine.

Most electrical circuits in a building carry 15 to 20 amps of electricity. Usually more than one audio/visual machine can be operated on one outlet. However, laminators and dry-mount presses must have one wall outlet each unless operated on a specially designed 30-amp circuit. You cannot tell a 15-amp from a 30-amp circuit with the naked eye. Someone familiar with your building must tell you.

All equipment that is meant to be used together should be plugged into the same outlet, even if you need a multiple outlet strip. Videotape equipment is an example. In using several outlets, the equipment may not work properly.

If the power is off on your equipment, the problem could be in the equipment or in the building. Try a light or minor appliance in the wall outlet. If it works, your equipment has the problem. If it does not work, the problem is with the building circuit; perhaps there are too many items on one ciruit.

Since there are no standards for wiring buildings, there could even be several rooms operating on the same circuit. Someone familiar with the building would know the wiring.

Below are some basic electrical terms:

Ampere. A measure of current flowing through a particular circuit. Each circuit can only carry a limited number of amperes. If that amount is exceeded the fuse will blow.

Circuit. A circular path or closed system allowing uninterrupted flow of electricity.

Voltage. The "push" or pressure of electrons needed to create electricity through a circuit.

Wattage. The amount of power needed to operate a particular piece of equipment (amperage multiplied by voltage).

The example below can help you figure out how to prevent an overloaded circuit: Say it takes 1,000 *watts* to operate a transparency maker. You are probably using 110 *volts*, and you've been told this circuit can handle up to 15 *amperes* of current.

Can you also plug a tape recorder into this circuit? First see how many amperes the transparency maker is using.

$$\text{Amps} = \frac{\text{watts}}{\text{volts}} \text{ or } \frac{1{,}000 \text{ watts}}{110 \text{ volts}} \text{ or about 9 amperes.}$$

Now check the wattage marked on your tape recorder and figure the amperage as explained above. Will you be using a total of 15 or less amps for both pieces of equipment? If so, under normal conditions you will not have an overloaded circuit.

WALL PLUGS AND CORDS

WALL PLUGS

- Make certain that the machine controls are OFF when plugging the cord into the outlet.

- Three-contact wall plugs (one contact with round ground) should fit three-contact grounded outlets. If you have a two-contact outlet, use an adaptor. The ground wire on the adaptor should be attached to the outlet wall plate. Otherwise, it will not serve the grounding function of preventing electrical shock. (See Figure 1-4.)

- With a two-blade plug where both blades are the same size, reverse the plug if there is a hum in the equipment or you feel a shock.
- Plugs with one large and one small blade must fit the wall outlet. There are no adaptors.
- Never force a plug into an outlet. Make sure the plug and the outlet are compatible. Get an adaptor if necessary.
- Unplug the machine from the wall when it is not to be used for a long time. This will prevent possible shock and fire hazard from lightning storms or power surges.
- To disconnect the cord, pull it by the plug, never by the cord.

CORDS

- Keep all cords in good repair to avoid shock and to provide effective operation of the machine.
- Keep AC cords well separated from equipment and other cables. If they are too close, there may be noise or other interference in the picture or sound.
- Three-pronged wall plugs on machine cords need three-holed extension cord receptacles. Do not use two-wire adaptors to connect the equipment plug to the extension cord. However, you may use an adaptor at the outlet.
- Do not use thin, light-duty extension cords for any projector or other A/V equipment, except cassette recorders. There are two reasons for this:

1. The cord gets hot quickly, posing a fire hazard.

2. Extension cords cut down on power to the machine—the machine does not operate as efficiently (slower, less brilliance).

WHAT YOU NEED TO KNOW ABOUT SHELF LIFE

The supplies listed below have a shelf life or use life after which they are no longer effective. High temperature and humidity shorten life.

Film (still or movie). Has the expiration date stamped on it.

Batteries. Last for a certain time even if unused. Test them if in doubt.

Lamps. Are good for a certain number of hours of usage. Moving equipment while bulbs are still hot will shorten their life considerably.

Thermal transparency film. Has a shelf life of one year or more, depending on the manufacturer and the quality. The dealer will know how recent the box is.

STORING EQUIPMENT

1. Replace lids, covers, and so on, when the equipment is not in use for a period of time.
2. Store equipment in a cool, dry place. Extremes in temperature, dampness, and dust will affect performance and may damage equipment. Never leave equipment on a radiator, on a car dash, or in a car trunk.
3. Keep equipment locked up in cabinets or closets, whenever possible.

HOW TO DETERMINE THE MODEL OF YOUR MACHINE

The entire model number consists of the printed number plus any additional letters or numbers engraved or embossed next to model number. For a particular model, these extra markings might denote a particular series. For example, Beseler Porta Scribe Overhead G–100 is the basic model that comes in XX, AA, SS, and RR series. This is particularly important when looking up a lamp code or when you need replacement parts.

CLEANING AND SERVICING EQUIPMENT

• Keeping equipment clean and having it serviced at least once a year are the greatest keys to dependable, trouble-free, longer-lasting equipment operation. You can do this yourself or if the equipment is complicated, as with a 16mm projector, take it to a reliable audio/visual service center.

• Generally, photographic and projector lenses (cameras, slide, filmstrip, movie projectors, overhead projector head) are to be wiped with lens tissue, not facial or eyeglass tissues. Other glass surfaces can be cleaned with clean, soft, lint-free, damp cloth.

• For cabinets, never use acetone or alcohol. Use a clean, soft, dry cloth. For all surfaces, always wipe very gently.

• Metal parts can be cleaned with cotton swabs and alcohol. Use 95 percent isopropyl alcohol. Ask your pharmacist or audio dealer for this. Isopropyl rubbing alcohol is inadequate because of the glycerine content, which leaves a residue on equipment after cleaning.

• Keep machines covered, with either a hard cover or a soft plastic cover.

• Do not leave machines near chalkboards; they will get covered with dust.

EQUIPMENT REPAIR

• Do not attempt to repair equipment yourself if it requires unscrewing the cover or working inside the machine.

• Do not oil machines yourself. Too much oil can damage equipment.

• If there is something wrong with the machine you are using and you cannot repair it, attach a note telling what the symptoms are and the possible cause of the problem, if you know it. This will help the repair service a great deal. In addition, the equipment will not be passed on to another user before it is fixed.

2 Lamps

CONTENTS

Operating Tips	10	**InDepth**	
Cooling Lamps	10	Determining the Proper Lamp	12
Inserting/Removing Lamps	10	Keeping the Same Lamp Code	12
Replacing Lamps	11	Using Codes in a Photographic Lamp Guide	12
Problems	11	Different Kinds of Lamps	12

Special lamps are used in audio/visual equipment. They must be purchased from an audio/visual dealer or camera store. Depending on the type of lamp needed, the retail price can range from less than $5 to $30 per lamp.

OPERATING TIPS

I. Cooling lamps

Cool the lamp before turning the equipment off and moving it, or before replacing the lamp.

A. Switch the machine to FAN, with the lamp off. Wait until the air from the projector is cool, and turn projector off.

B. Where there is no fan, turn machine off, but do not move the projector until it has cooled.

II. Inserting and removing lamps

A. Unplug the equipment cord before inserting the new lamp.

Examine the socket and lamp base before inserting the lamp into the machine. The lamp will have a shorter life if not fully inserted into socket.

Keep fingerprints off the lamp. Use paper or cloth to hold the lamp when removing it from the box or installing it. If you do touch it, gently wipe glass with isopropyl alcohol and lint-free cloth or paper towel wet with soap and water. Wipe with paper towel moistened in fresh water, then pat dry.

B. To remove the lamp, screw off or snap off the lamp covers.

Then, determine whether (1) there is a lamp ejector lever, and (2) whether the lamp must be turned in the socket before pulling it out.

If it cannot be removed easily, wiggle it gently and pull at the same time.

III. Replacing lamps

Replace a lamp with one of the same lamp code and wattage as called for by the manufacturer.

PROBLEMS

PROBLEM	CAUSE	SOLUTION
1. Lamp burns out.	Old lamp.	Replace.
	Air in lamp. Note gray-white film on inside of lamp.	Replace.
	Lamp not placed properly in socket.	Note how lamp is inserted—some must be pushed in and turned.
	Lamp not cool when projector was moved.	Switch projector motor setting to FAN. Run projector on FAN until air from machine is cool. Then turn to OFF. Where there is no FAN switch, turn projector off and allow lamp to cool.
	When using HIGH/LOW switch, HIGH is used too often.	Use LOW switch.
	Lamp touched with hands.	Never touch lamp (see Operating Tips II).
2. Lamp melts or damages projector cabinet.	Used lamp with wrong letter code, though it may fit socket.	Check serial plate on projector for lamp code.
3. Poor illumination.	Dark-looking lamp.	Replace lamp.
	Lamp with built-in reflector not in proper position.	Check lamp to see if it is to be pushed in or turned to final position.
4. Gray-white film on inside of lamp.	Air inside lamp.	Replace lamp.

IN-DEPTH

I. **Determining the proper lamp for your equipment**

   A. Each lamp is identified by three letters, called a lamp code.
      1. The equipment specifies the proper lamp code on its serial plate.
      2. The lamp itself has the code stamped onto the glass or base.
      3. Your dealer will know the lamp code if you give the make and model of your machine as noted on the serial plate. Include printed, engraved, and stamped numbers *and* letters. Some manufacturers use a combination of these three ways to express the model.
      4. Obtain a General Electric or Sylvania *Photographic Lamp and Equipment Guide.* It gives the lamp code for your machine by brand and model. It is free and is updated every few years. Locally, call the General Electric Lamp Division, or write to:

         General Electric
         Lamp Business Group
         Nela Park
         Cleveland, Ohio 44112

         For Sylvania lamps, write to:

         GTE Products Corporation
         Lighting Center
         Danvers, Massachusetts 01923

II. **Keeping the same lamp code**

   Generally, lamps are *not* interchangeable. The lamp code remains the same for the make and model of equipment regardless of who manufactures the lamp. For example, a Beseler overhead projector VU-graph Master 6600 *always* takes a DRB or DRS lamp—whether General Electric or Sylvania makes it.

   Do not substitute a different lamp code, even if the lamp looks the same and fits the lamp socket. Do not substitute wattage. Use the lamp specified by manufacturer.

III. **Using codes in a photographic lamp guide**

   A. CYR/CYL—A slash between codes indicates *one* lamp code only. The equipment serial plate may have one or both sets of letters, but the lamp will carry both. Order by the first set of letters.

   B. CYC, CTM—A comma between codes indicates *two* lamp codes. Either lamp code may be used for your piece of equipment.

IV. **Different kinds of lamps**

   Usually, you need only a projection lamp for a given machine. However, 16mm movie projectors with optical sound track (the kind commonly used in schools) take two or three lamps: a projection lamp, a sound (exciter) lamp, and sometimes a threading lamp. Photographic lamp guides list all three lamps under the make and model of the projector. You need a lamp code for each type of lamp.

3 Overhead Projectors

CONTENTS

Operating Tips 13
Problems 14
InDepth
    Screen and Projector Relation 15
    OFF-ON-Type Switches 15
    Overhead Techniques 16
    Cleaning Your Overhead 18
Glossary 18

The overhead projector is one of the easiest projectors to operate and can be used for a variety of teaching techniques. It is fairly inexpensive. A beam of light shines through transparent pieces of acetate (transparencies), and the image is then projected on a screen.

Transparency presentations are available commercially. Often, however, they are made by hand in one of two ways:

1. Writing or drawing directly on the acetate.

2. Transferring print or images done on paper onto special, clear thermal film using a transparency maker.

The overhead projector has several advantages:

1. It can be used in a lighted room, allowing students to take notes.

2. The lecturer is facing the audience and has eye-to-eye contact with individuals.

3. It can replace a blackboard. Repetitive material can be shown several times.

4. It takes up little space but can project images as large as a blackboard can.

5. Color can be used effectively.

The most critical factors affecting overhead projection are the placement of both the screen and the projector (see InDepth I).

OPERATING TIPS

1. Do not leave the projector in a car or another place where the sun might shine on it directly.

2. Never carry the projector by the projector arm or post. Always carry it by the base. The post will eventually break or become detached from the projector.

3. To write on acetate used for overheads, use pens and pencils designed for overhead

PROBLEMS

PROBLEM	CAUSE	SOLUTION
1. No light on screen.	Power cord not plugged in.	Check power cord connection.
	Opaque material on screen.	Use clear acetate.
	Light switch OFF or not turned all the way ON.	Turn switch ON.
	Lamp burned out.	Replace lamp.
	Suspect lamp socket burned out.	Take machine in for repair.
2. Not enough light on screen.	Dusty fresnel lens and/or top glass.	Clean with damp, lint-free non-abrasive cloth and very mild dishwashing detergent. (See InDepth IV.)
	Old, darkened lamp.	Replace lamp.
	Thin extension cord.	Use only heavy-duty cord.
3. Uneven illumination or colored areas in corners of screen.	Optical system out of alignment from rough handling.	On some 3M machines, adjust lamp focus. Otherwise, take in for repair.
4. Picture not square on screen—top or bottom or one side larger than the other (keystoning).	Projection or screen angle needs adjustment.	Light beam from front lens of projection head should be perpendicular to center of screen.
5. Picture out of focus.	Focus adjustment necessary.	Use focus knob.
	Projection arm out of line or broken.	Take in for repair.
6. Dark spot, out of focus.	Fresnel lens damaged or warped.	Replace lens. Make sure fan is working, where available.
7. Fan keeps running after switch turned to OFF.	Fan is cooling lamp.	On newer equipment, fan will automatically cut off when lamp has cooled.

projection; otherwise, the writing will break up on the acetate, and colored pens or pencils will write black. Pens can be alcohol-based, water-soluble, or permanent felt-tip. The advantage of alcohol-based pens is no smearing on the fingers while writing, but the writing can be erased easily using a cloth moistened with alcohol.

4. When making transparencies, use large type with double or triple spacing.

5. If the top glass of the projector is cracked, do not remove it until a replacement arrives. The lens under it (fresnel lens) is made of plastic and scratches easily.

6. Shut the door to the lamp compartment. This keeps the lamp cooler and keeps dirt and other foreign matter out.

7. Turn off the lamp if the cooling fan does not come on after two to three minutes. This will prevent heat from melting the fresnel lens.

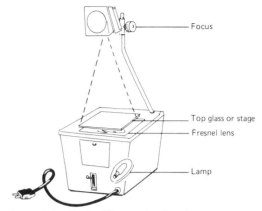

Figure 3-1 *Parts of the overhead projector.*

IN-DEPTH

I. **Screen and projector relation (see Figure 3-2)**

  A. Location of screen—For the maximum viewing audience, place the screen in the front corner of the room on the window side.

  B. Screen and projector angle—The projector head beam and the screen must be perpendicular to each other (90-degree angle to each other). Move the angle of the screen; move the angle of the projector head, or both.

II. OFF/ON-**type switches**

  It is very important that the projector and the lamp be allowed to cool after the presentation; otherwise, the fresnel lens can bubble, which will affect focus. Fresnels are expensive.

  There are three types of lamp switches. Find the switch below that applies to your projector.

  A. OFF/FAN/LAMP switch—When you are finished, turn the switch to FAN to let the lamp cool. Do not immediately turn switch to OFF.

  B. OFF/ON switch (without automatic fan)—When you turn the machine off, the fan also goes out. The machine will not cool itself. Do not move the machine until the lamp and the projector are cool.

  C. OFF/ON switch (with automatic fan)—When you turn the machine off, the fan keeps running until the lamp cools. Then the machine and the fan will turn off automatically. Some projectors use numbers on the switch, such as 1 for ON and 0 for OFF.

Figure 3-2 *Projector-screen relation.*

III. Overhead techniques

Below is a list of things you can do with your overhead to provide variety and interest:

A. Use your overhead projector like a chalkboard. While talking to your audience, explain your ideas by writing or drawing on a blank acetate. For a predrawn acetate, use a pointer to identify specific information.

B. Use overlays to help tell a story with scenes building on each other (see Figure 3–3).

C. Use the revelation technique by placing a piece of paper on the stage to block out all but the first portions of your message. Then, slide the paper down to reveal more information. Do not block the light for too long, or the top glass may heat up and crack.

D. Use the ON/OFF projector switch to have the audience concentrate on

Figure 3-3 *A story on overlays—first, draw the entire picture on paper; second, in your mind divide the story into beginning, middle and end. Trace each part from paper onto a separate acetate sheet. Third, project acetate 1. Tell your story. On the projector stack acetate 2 on top of acetate 1. Continue your story. Repeat until you have used all acetates and finished your story.*

you, rather than on visuals, when necessary.

E. Show translucent models to explain the interaction of moving parts. An opaque model will show up in silhouette form.

F. In preparing your acetate, add color by:
   1. Using a colored piece of acetate as your original.
   2. Writing with appropriate colored pens and pencils.
   3. Adding pieces of colored transparent adhesive film to the original clear acetate (available at art supply stores).
   4. Using India ink on special acetate.

G. Make paper copies from your transparency, to be used as handouts.

H. Letter the acetate with dry transfer lettering (available at art and school supply stores). You will get professional-looking results with little effort.

I. Use artist's tape to draw lines. Various solid colors, patterns, and widths are available in transparent or opaque tape at art supply stores.

IV. Cleaning your overhead

The top glass and fresnel lens need frequent cleaning. Dirt from these surfaces appears enlarged on the screen. Wipe both the fresnel lens and the top glass with a lint-free cloth dampened with 95 percent pure isopropyl alcohol. Instead, you may dampen the cloth with a very mild dishwashing detergent. Make sure to rinse away the detergent. Dry gently with another soft, lint-free cloth.

GLOSSARY

Acetate Transparent material on which information is written or drawn for overhead projection.

Fresnel lens Main lens for the projector, immediately under the top layer of glass. The lens is plastic, scratches easily, and is quite expensive. Use care in cleaning.

Platen *See* Top glass.

Stage Platform upon which the transparency is placed.

Top glass External glass covering the fresnel lens. Place your material directly on the top glass. Be careful in applying pressure, such as when writing, because this glass is thin and breakable. Also called *stage glass* or *platen*.

4 Opaque Projectors

CONTENTS

Operating Tips	19
Problems	20
InDepth	
Operating Procedure	20
Cleaning	21
Glossary	21

The opaque projector is a useful though bulky machine that projects and enlarges images on a screen. It can handle opaque objects such as art prints, pages from a book, coins, shells, and other opaque objects. It can handle objects 10 inches long by 10 inches wide that are up to 2 inches thick. It must be used in a darkened room.

OPERATING TIPS

1. Establish projector–screen distance by focusing your largest picture on the screen before beginning the program. This projector-to-screen distance will be applicable to most smaller pictures you will be using that day.

2. Operate in total darkness.

3. Before inserting material, make sure that the heat-absorbing glass pressure plate is present by touching the area above the platen. This glass protects your material. You may, however, need to remove it when using three-dimensional objects.

4. Attach small pictures and clippings to heavy cardboard to keep them flat.

5. Make sure the fan is working; otherwise, the material may scorch and curl.

6. Do not insert high-gloss materials, including photographic materials such as black and white or color prints; they may fade, curl, or bubble within fifteen seconds.

7. Do not open the platen when the projector lamp is on. This can distract the audience.

8. Use the projector to help draw bulletin board art and posters. Tape your drawing paper to the wall. Insert the original drawing in the projector. Determine the size of your enlargement by moving the machine closer or further away from the paper. Focus the projector and trace the enlarged projected image using pencil, pens, or markers.

PROBLEMS

PROBLEM	CAUSE	SOLUTION
1. Image will not focus, even with lens out all the way.	Projector too far from screen.	Move projector closer.
	Material not held flat in projector.	Adjust height of platen.
2. Material scorches and curls, fades, or turns yellow.	Fan not working properly.	Take in for repair.
	Heat-absorbing glass pressure plate missing.	Contact audio/visual equipment dealer.
	High-gloss materials used.	Avoid high-gloss subjects.
	Incorrect lamp used.	Use only lamp recommended by manufacturer. (See Lamps).
3. Built-in pointer not working.	Pointer lamp not working—broken cable.	Take in for repair.
4. Poor illumination.	Dirty lens.	Wipe lens.
	Room not dark enough.	Darken room, or move projector closer to screen.
	Old, darkened lamp.	Replace lamp.
	Using thin extension cord.	Use heavy-duty cord.
	Broken mirror.	Take in for repair.
5. Lamp lasts for only three to five hours of use.	Lamp not making good connection in socket.	Clean center contact on lamp socket with pencil eraser. If contact is pitted, use fine emery cloth.
	Bad switch.	Replace switch.

IN-DEPTH

I. Operating procedure

A. Slide the platen open with the lever.

B. Insert material to be projected. If using a printed page, the print should be inserted upside-down.

C. Make sure that the heat-absorbing glass pressure plate is over the material, unless you have coins or shells, or other 3-D objects.

D. Close the platen so that the object is somewhat snug between the glass pressure plate above, and the platen

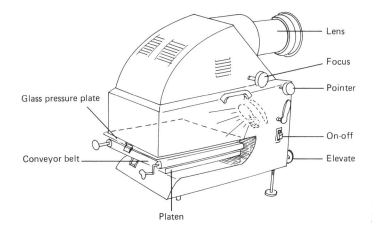

Figure 4-1 *Parts of the opaque projector.*

below. This is particularly useful for keeping an open book flat.

E. Turn the lamp switch to LAMP so that both the fan and the lamp are operating.

F. Focus the image.

G. Use the built-in pointer, if available and necessary.

H. After finishing, turn the lamp switch to FAN to allow the projector to cool. If there is no FAN switch, do not move the machine until it is cool.

I. Remove the material from the projector.

II. Cleaning

This projector requires little maintenance. Lenses should be cleaned with a soft, dry cloth. If foreign matter or dust gets on the mirror inside the projector, just blow it off. The mirror scratches easily if touched.

GLOSSARY

Platen Surface of the projector on which the material to be projected is placed.

Stage *See* Platen.

Pointer Movable lighted area controlled by a knob near the focusing knob. Many models have this feature.

5 Dry-Mount Press

CONTENTS

Operating Tips	22
Problems	23
InDepth	
Temperature Settings	24
Procedure for Dry Mounting	24
Cleaning the Machine	25
Glossary	26

The dry-mount press is a versatile piece of equipment that joins materials together through heat, pressure, and a heat-sensitive binding agent instead of ordinary paste, glue, or cement. The press comes in several sizes from 8½ by 11½ inches to 26 by 32 inches. You can perform four functions with the press:

1. Mount material on a firm backing—useful for photographs, painting, fabrics, and so on.

2. Mount material on a flexible cloth backing to keep it pliable but tear-proof—useful for maps and charts that you want to roll.

3. Laminate—especially useful for thick materials, such as plywood or masonite, which do not fit into an automatic laminating machine (see Glossary—Lamination).

4. Make transparencies through a heat-lifting process (see Glossary—Heat lifting) for overhead projector use.

The press has a thermostat control to set the temperature for each function.

OPERATING TIPS

1. Before closing the press, cover the material with a heavy sheet of paper larger than the original so that no ridge marks will appear on the finished product. Blank newsprint provides good protection.

2. For thick materials, know the maximum thickness of material that your machine will accept. Usually, presses can dry mount fairly thick boards.

3. For thin materials, place a few sheets of cardboard in the press beneath the pad on which you place the material.

4. For wide and long materials, use materials up to twice as wide as your press and of indefinite length. For long materials, start at the center and work toward either end. For wide and long materials, tack the material to dry-mount tissue, but do not tack it to the mounting board before dry mounting. This will help to prevent wrinkles on the final product.

PROBLEMS

PROBLEM	CAUSE	SOLUTION
1. Air bubbles.	Moisture in either material or mounting board.	Wrap mounting board and material in brown wrapping-paper or art paper envelope. Place both mounting board and material in press for 30 seconds at 225 degrees. Remove, and prepare for mounting.
	Thick material.	Keep in press longer.
	Not enough pressure.	Add some cardboard backing. Dry mount again.
2. Wrinkles during dry-mount press lamination.	Wrinkles *cannot* be eliminated.	Use solutions below to prevent future wrinkles.
	Film tacked to material before laminating.	Do not tack film to material before laminating.
	Insufficient heat.	Leave in press long enough.
	Surface of material not smoothed as it is placed in press.	Laminate one side at a time. Smooth material before mounting. Wrap in brown paper envelope before laminating.
	Material being laminated is soft.	Put cardboard under dry-mount press pad on which you lay material.
3. Curling or warping of finished product.	Moist when put in press.	Before mounting, dry material in press for 30 seconds at 225 degrees.
	Need pressure on finished product.	After mounting, place heavy book on material until cool.

IN-DEPTH

I. Temperature settings (see Table 5-1)

It takes about 11 minutes to warm up the press to 225 degrees.

Times and temperatures given in Table 5-1 are approximate. Initially, use the minimum time and temperature necessary. You can always increase these or return your material to the press for a better job.

The thermostats are not always accurate. Check the press temperature by placing a square metal thermometer at the edge of the lid when shut. A candy thermometer will do fine, or use the sizzle test: With your moistened finger, touch the platen. A slight sizzle indicates just right for dry mount. No sizzle tells you to let the machine get warmer. A loud sizzle says to allow it to cool down for dry mount.

Table 5–1. Dry mount temperature/time guide.

FUNCTION	TEMPERATURE SETTING	BACKING MATERIAL OR TISSUE	APPROXIMATE TIME IN PRESS
Dry mount	225°–250°	Dry-mount tissue and mounting board (see Glossary for Chartex and Fotoflat)	30–45 seconds
Laminating	270°–350°	Laminating film	1 minute +
Heat lifting	270°–350°	Laminating film	1 minute +

Tacking iron—If it has temperature settings, usually place on MEDIUM. It should be hot to the touch before using. Alternative to tacking iron: tip of clothes iron set on DELICATE, not on steam.

II. Procedure for dry mounting

Equipment and materials: dry-mount press, tacking or clothes iron, scissors and/or paper cutter, your material, mounting board (such as poster board), dry-mount tissue, envelope of blank newsprint, brown paper, or art paper.

A. Select your material. Do not trim.

B. Preheat the dry-mount press and the tacking iron.

C. Put the material face down. Place the dry-mount tissue on the back of your material.

D. Use the tip of the iron to tack the tissue to the material. Tack at the center and three-quarters of the way toward each corner. Think of a small × pattern. Do not tack at the corners (see Figure 5–1a).

E. Trim the material and the tissue to the same size (see Figure 5–1b).

F. Place the material on top of the mounting board. Center or position the material. Separate the picture from the loose tissue at the corner. Place the corner of the tissue against

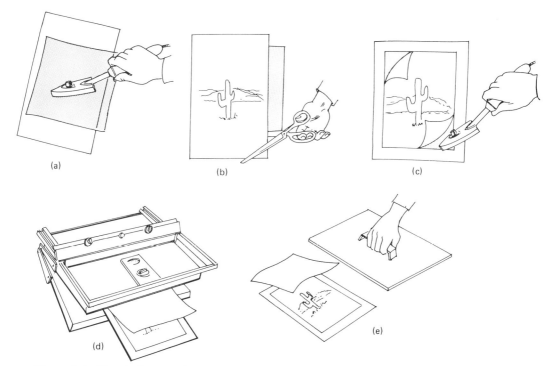

Figure 5-1 *How to dry mount.*

the mounting board while holding the picture in place. Touch the iron to the tissue so it will adhere to the mounting board. Repeat at one other corner. Omit this step for extra wide and long materials (see Figure 5–1c).

G. Wrap the material and the mounting board in blank newsprint, brown paper, or art paper envelope to protect it. You may also use release paper, which is a commercially available product. Place the waxed side next to your material.

H. Place in the dry-mount press. The material to be dry mounted should face up. Keep in the press at 225 degrees for 30 to 45 seconds (see Figure 5–1d).

I. Remove and inspect the product for bubbles. Insert again if more heat is needed.

J. Place a heavy weight on the mounted material until cool (see Figure 5–1e).

K. The material is now ready for use.

L. Turn the dry-mount press off. Unplug the tacking iron.

III. Cleaning the machine

For stubborn deposits on the platen, use fine-grade emery cloth available at a hardware store, fine-grade (00 or 000) steel wool, or a razor blade. Wipe clean with soft damp cloth. Avoid marring the platen.

GLOSSARY

Chartex Cloth backing (the side without cloth is the adhesive). Use the dry-mount press at 225 degrees for 5 seconds.

Dry-mounting tissue Special waxy tissue paper coated on both sides with a heat-sensitive adhesive. When sandwiched between the print and the mounting material, it forms a bond between the two.

Fotoflat tissue Dry-mounting tissue that can be removed at a later date so that the original print can be separated from its backing. To dry mount, use the press at 180 degrees. To remove the backing, reheat the material at 200 degrees for one minute, then use a razor blade to separate the material carefully from the mounting board.

Heat Lifting A process for making transparencies from paper originals for use with the overhead projector. It is especially helpful for converting magazine pictures into transparencies. Generally, use a magazine page that is clay-coated. Laminate only on the side of the desired picture, not on both sides. Soak laminated product in soapy water. Gently separate or rub away paper from film. The image will remain on the laminating film, and this film is your transparency.

Lamination Heat and pressure process of encasing and sealing material between two layers of plastic film.

Platen Metal surface on the inside cover of a dry-mount press, which conducts heat to your material.

Tacking iron Essential accessory to dry mounting. This small iron is used to attach (tack) mounting tissue to your material and to the mounting board before they are put into the press. It is also useful in taking out air bubbles left on the finished product. (If a tacking iron is not available, use the tip of a clothes iron, not on steam setting.)

Tear sheet A term for the material you have selected to be dry mounted.

6 Laminators

CONTENTS

Operating Tips 28
Problems 28
    Product 28
    Machine 32
InDepth
    Major Parts of the Laminator 34

Operating Laminator 35
Using New Film 36
Cleaning the Machine 40
Glossary 40

Laminating is the process of encasing and sealing material in plastic film. Heat and pressure are used in this process. The plastic film protects your material from damage and deterioration.

Many kinds of materials—from onionskin to cardboard, wood, and cloth—can be laminated. The rougher the surface, the better the lamination. Those materials that do *not* laminate well permanently are:

- High-enamel papers such as those used in many magazines
- Other papers with a slick finish
- Smooth metallic, plastic, chalk-covered, or crayon surfaces
- Some photographic paper (it may scorch or melt)
- Moist materials (they cause bubbles on the laminated surface)

Laminating machine features are as follows:

1. Machines come in various widths, from 9 inches to 60 inches wide; the average is 12 inches.

2. Most machines will laminate only light materials, while others will accept a range of materials from a few thousandths of an inch to an eighth of an inch thick.

3. Machines come with one speed or with variable speed. The variable speed adjustment allows heavy materials to be fed in more slowly than light ones.

4. Know your machine. Label the various parts and their functions on the machine itself.

OPERATING TIPS

1. Give yourself plenty of time when laminating.

2. Use the OFF button in case of problems. Do not panic. Many laminating problems can be solved with patience and careful handling.

3. Have available an Exacto knife and fine-grained copper wool pad in case of problems.

4. To prevent jamming of film, when first feeding the material, be prepared to grasp immediately and firmly take hold of the laminated product as it exits the machine. This is very important.

PROBLEMS

PROBLEM	CAUSE	SOLUTION
Product Problems		
1. Blisters, bubbles, or one large area that has not adhered properly.	Machine too cool.	Let machine warm longer. Remove film from material if possible. Relaminate.
	Dirty laminating rollers.	Clean with rubber eraser, or lightly rub with fine-grained copper wool pad. Relaminate.
	Excessive moisture in material.	Reduce moisture before laminating: place in warmed dry-mount press at 225 degrees for 30 seconds. After laminating: prick bubbles with pin; then seal film with tacking iron or, for large areas, place in dry-mount press (see Figure 6–1).
	Material slick, or powdery surface (chalk or pastels).	Will never laminate well.
2. Wrinkles	Wrinkles *cannot* be eliminated.	
	Need machine adjustment for wrong tension, too much heat, nonfunctioning cooling fan, or wrong pressure for laminating and pull rollers.	Remove film from your material. Make machine adjustments where possible, or take in for repair.* Relaminate.

*Make adjustments only if the knob or switch is easily accessible. Do not touch the inside of the machine.

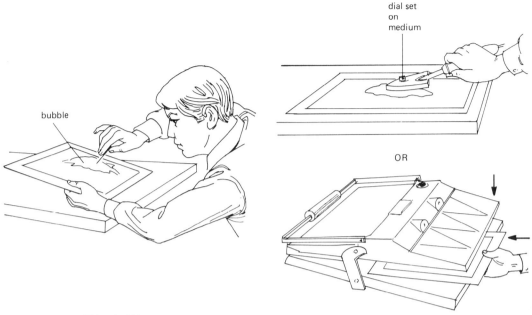

Figure 6-1 *Bubbles, bubbles, go away...*

	Dirty laminating rollers.	Clean with rubber eraser, or rub lightly with fine-grained copper wool pad.
	Tried to reposition material after it was fed into machine.	Material cannot be repositioned once it has been engaged by laminating rollers.
	Forced material into machine.	Feed only at motor take-up speed.
	Wrinkles appear at glued areas when lightweight glued to heavyweight material.	Glue only at leading edge.
3. Poor seal at edges.	Adjust machine for temperature too cool, speed too fast, or laminating roller pressure on variable speed models.	Make necessary adjustments where possible, or take in for repair.* Then, relaminate, or use tacking iron to seal edges, put in dry-mount press at 270 degrees to 350 degrees.

*Make adjustments only if the knob or switch is easily accessible. Do not touch the inside of the machine.

30 Laminators

PROBLEM	CAUSE	SOLUTION
Product Problems	Material wider than your laminating film.	Trim material and relaminate, or use wider laminator with wider film, or see problem 4 below.
4. Laminating oversized materials.	Material too large for width of machine.	Method 1: You may laminate material twice the size of the machine. Fold material in half. Place it on feed table close to edge of film. Laminate. Trim three sides. Do not trim folded side. Refold your material inside out and repeat as above. When you open the finished product, tack down any loose film on fold line using tacking iron.

Figure 6-2 *Oversized materials can be laminated.*

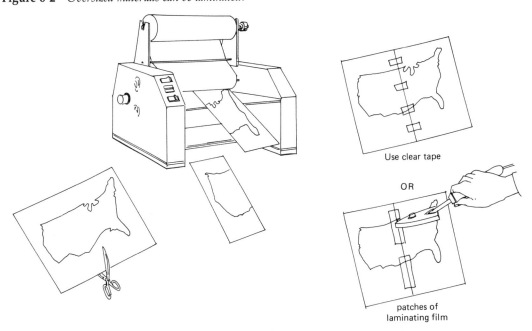

Method 2: (see Figure 6–2) Cut the material into major pieces. Laminate each piece separately. Align laminated parts to restore original figure. Patch parts together with clear tape, or cut pieces of laminating film to patch parts together. Use tacking iron or clothes iron set on DELICATE to laminate patches in place.

5. Two separate items accidently overlap and are laminated together (see Figure 6–3).	Materials were too close to each other when fed into machine.	Use Exacto knife to cut through film overlap front and back. Separate materials gently. Relaminate individually, leaving at least a half-inch of clear film between items.
6. Curling of finished lamination.	No heavy weight placed on finished lamination.	After using solutions to problems following, place laminated material under heavy weight until cool.

Figure 6-3 *Laminated materials that accidentally overlapped can be separated and relaminated.*

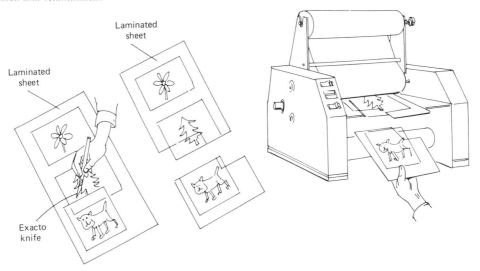

32 *Laminators*

PROBLEM	CAUSE	SOLUTION
Product Problems		
	Unbalanced tension in pull rollers or supply rollers.	Make necessary machine adjustments where possible or take in for repair.* Then relaminate the material upside down.
	Uneven heat.	Make necessary machine adjustments.* Then relaminate upside down.
	Where available, heating shoes not clean.	Clean the material gently with fine-grained copper wool pad. Relaminate.
	Excessive moisture in material.	Place the material in warmed dry-mount press at 225 degrees for 30 seconds. Then remove and place heavy book on top. Relaminate.
Machine Problems		
7. No power.	Machine not plugged in.	Plug machine into AC outlet.
	Blown fuse in machine.	Check power in wall outlet. Replace machine fuse. Fuse is usually found on back of machine or where rollers are located.
8. Laminating film jams in machine (*most common problem*).	Failure to grasp film when it first emerges from machine.	Unjam roller.**
	Failure of laminated material to clear exit path.	Unjam roller.**

*Make adjustments only if the knob or switch is easily accessible. Do not touch the inside of the machine.
**See note on page 33.

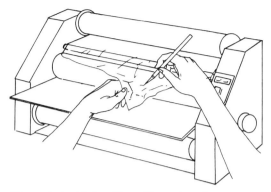

Figure 6-4 *Unjam film with CUT/RUN/PREHEAT technique.*

	One laminating roller without film, and the other roller winds film around itself.	Unjam roller.** Add new film.
9. Gummy substance accumulates on laminating rollers.	Poor alignment of top and bottom laminating film. Heat melts adhesive of overlapping film onto rollers.	Remove gummy polyethylene from roller with rubber eraser, or lightly rub with fine copper wool pad. Align top and botton film of supply rolls parallel to each other and matching end to end.

**Unjamming roller: Press OFF. Don't panic. Work slowly and patiently. Determine whether laminating rollers or pull rollers are jammed with film. On front of machine, cut away film into machine from both top and bottom supply rollers. Remove jammed film as follows:

FIRST TRY: With machine OFF, grasp loose ends of film at roller end and pull with force!

SECOND TRY: With machine on RUN, grasp loose ends of film, removing as much film as possible while the rollers turn.

LAST RESORT: Cut/run/preheat technique (see Figure 6–4):

1. Using Exacto knife, carefully cut film from laminating roller, *without touching roller.*
2. Detach film from roller.
3. If possible, holding film with one hand at exit, *briefly* press RUN, remove as much film as you can, then press PREHEAT to stop machine.
4. Repeat procedures 1 through 3 as many times as necessary.
5. Before reusing machine, extract all jammed film and press PREHEAT to warm machine.

IN-DEPTH

I. Major parts of the laminator

A. A laminator has a feed table and two sets of each of the following parts—one set for top and one set for bottom lamination.

B. Supply rolls—contain new laminating film.

C. Guide bars—keep the film flowing properly from the supply rolls to the laminating rollers.

D. Laminating rollers—provide the laminating process.

E. Pull rollers—pull the laminated product away from the laminating rollers in order for product to exit machine (see Figure 6–5).

Figure 6-5 *Parts of the laminator.*

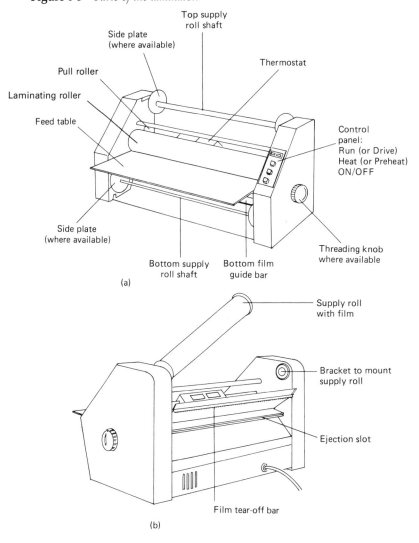

II. Operating laminator
   A. Before laminating your material:
      1. Check that the material is smaller than the laminator, to allow for proper seal around the edges.
      2. If the material is damp, or if it is a damp day, dry the material before laminating. Wrap the material in an envelope of brown wrapping paper, art paper, or blank newsprint. Insert in a dry-mount press at 225 degrees for 30 seconds. Instead of a press, you may substitute a variable speed laminator set at slowest speed.
      3. Always run a sample of your type of material. If you change material, run another test. If it does not feed easily, it is too thick.
   B. Setting up the laminator:
      1. Connect the power cord.
      2. Press PREHEAT or HEAT to warm the unit.
      3. Depending on the machine, the light will either come on or go out when the machine is ready.
   C. Feeding material in the laminator:
      1. Press RUN or DRIVE to start the laminating rollers.
      2. *To prevent jamming*, when first feeding material, feed with one hand and hold the other hand at the exit to grasp the material as it comes out. When it exits the laminator, it will be completely sealed.
      3. Feed material continuously at the laminator speed. Do not push. Leave a half-inch between what was just inserted and the next material to be laminated.
      4. If necessary, smooth out the surface of the material, working from the center toward the edges closest to you.
      5. When the material is engaged by the rollers, you cannot reposition the material on the feeding table. Wrinkles will occur if any changes are made.
      6. Do not stop the machine while the material is engaged by the laminating rollers; otherwise, a pressure mark on the finished product will occur at the stopping point.
   D. For temporary stop:
      1. Before stopping, make sure that the material is out of the machine and past the tear-off bar.
      2. While still on RUN, insert a thin cardboard until it is halfway between the entrance and exit slots.
      3. Press PREHEAT, HEAT, or STOP, whichever is applicable, to stop the machine.
      4. Do not let the machine stand for thirty minutes without using it. Switch it off.
      5. On some units, wait two minutes to press RUN or DRIVE if the machine is turned OFF and back on quickly.
   E. After laminating has been completed:
      1. Press OFF.

2. Cut off the material, leaving a few inches of film past the exit slot.
3. Trim the edges of the product with scissors or paper cutter, leaving an eighth-inch margin of laminated film all around for total seal.

III. **Using new film**
   A. Ordering film—four factors are important:
      1. Thickness of film (average is 0.0015 millimeters).
      2. Width of film (should be smaller than shaft width).
      3. Diameter of film core (sizes are 1 inch and 2¼ inches; average is 1 inch).
      4. Manner in which the film is wound on the roll (determine whether the shiny or dull side of the film shows on top of the old roll before discarding it).
   B. Knowing when film is running out:

   To alert you, a paper label appears on film within 3 feet of the end of the supply roll. Either use the film until the end, or cut and replace the old rolls at that point. Save end-of-roll remnants. They can be used for patching together oversized materials that must be laminated in sections.
   C. Loading film (see Figure 6–6):
   NOTE: Both supply rolls of film should be changed at the same time, even if only one roll is empty.

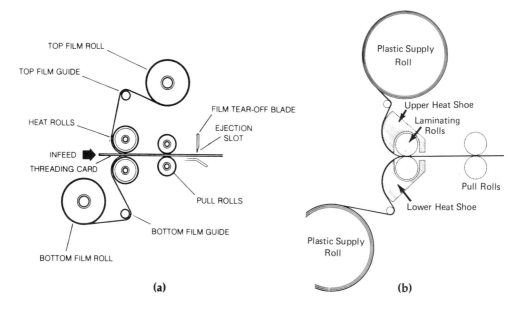

Figure 6-6 *Two kinds of threading procedures. (a. Courtesy General Binding Corporation, Northbrook, Illinois; b. Courtesy Laminex Industries, Inc.)*

1. To remove old supply rolls, turn the machine to PREHEAT or HEAT.

   Remove the feed table from the laminator.

   Remove the empty supply rolls from the machine. If one roll still has film feeding into the machine, use scissors or an Exacto knife to cut the remaining film from that supply roll. Remove the roll.

   If there is film already on the laminating rollers, allow it to run through the machine.

   Hold onto the film as it exits the machine. Press RUN.

   Press OFF when all of the film is out of the machine.

   Where applicable, remove the metal side plates on both supply rolls.

   Remove the shaft from both supply roll cores.

2. When inserting new supply rolls and threading, the laminator should be OFF.

   Supply rolls: Lay the new supply rolls out in front of the machine. Unwind some film from each. Note that once the supply rolls are on the machine, the dull sides of the film must meet where the laminating rollers meet. If you are not sure which is the dull adhesive side, place the roll on a table. The film will curl toward the adhesive side.

   Shafts: Check the way the shafts fit into the machine. They will only go in one way.

   Insert a shaft into each new supply roll.

   Where applicable, replace the metal side plates after the supply rolls are on the shafts. If the plates fit too loosely, they may be inside out. Replace a plate on the thin end of the shaft first.

   Replace each shaft with a roll on the machine.

   Check that the top and bottom supply rolls are exactly aligned with each other, matching end to end. They must also be aligned with the laminating rollers. Check that the dull sides of the film will meet at the laminating rollers.

   Guide bars: Depending on your model, the film should go over or under the guide bars, actually touching the guide bar when ready to feed into the laminating rollers. There is a top guide bar and a bottom guide bar; the bottom guide bar is sometimes hard to locate.

   Patience will triumph.

D. Engaging the film in machine before replacing the feed table (see Figure 6–7):
   1. Unwind some film from both the top and bottom rollers.
   2. Overlap the film so that the edge of the film for the top supply roll extends almost to the bottom supply roll, and vice versa. Carefully overlap and wrap both film layers inside each other a couple of times.

38 Laminators

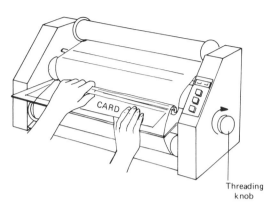

Figure 6-7 *Overlap film and insert cardboard between rollers.*

3. Hold the wrapped-up film with your left hand where the laminating rollers meet.

4. With your right hand, insert and push a thin sheet of cardboard against the overlapping films and into the laminating rollers. This forces the overlapping films between the rollers.

5. If your laminator has a THREADING KNOB, usually found on the right side of the machine, turn it clockwise until the cardboard is gobbled up by the laminating rollers. This action sometimes

Figure 6-8 *Clean your laminator. (Courtesy Laminex Industries, Inc.)*

requires brute strength. When the film is engaged, let the knob go.

6. Press PREHEAT or HEAT.
7. Press RUN when the machine is ready.
8. Grasp the cardboard *quickly* as it exits the machine. Press OFF.

   NOTE: If the cardboard does not exit, the THREADING KNOB, where available, is not turned enough, or one laminating roller may have the overlapped film wrapped around itself. Cut the film from that roller, pull out more film. Put it where the rollers meet at the front of the machine. Press run. When the cardboard exits, be ready to grasp it quickly. Immediately press OFF.

9. Press PREHEAT or HEAT, and allow the machine to warm up.
10. If there are wrinkles on the film of the top laminating roller, center the top supply roll on the machine.

Figure 6-8 (continued)

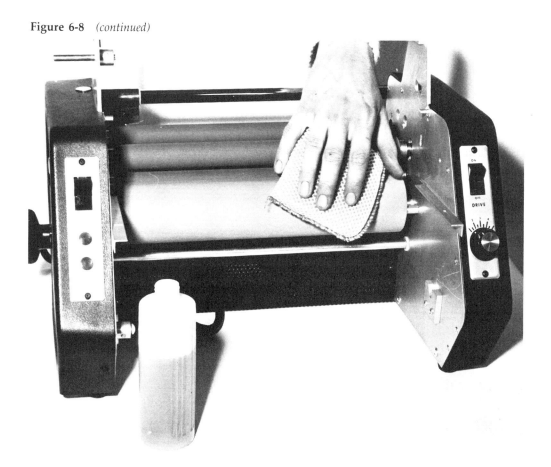

11. Press RUN or DRIVE to test that the film is laminating properly and that it is wrinkle-free.
12. Cut the film-encased cardboard, leaving a few inches of film at the exit point.
13. Install the feed table.
14. The machine is now ready to laminate your material.

IV. **Cleaning the machine**

Clean while the machine is warm but *unplugged* (see Figure 6–8, pages 38 and 39).

A. Laminating and pull rollers:
1. For light deposit of gummy polyethylene substance, use an ordinary rubber eraser, or use a special polyethylene remover available from your dealer. Wipe on with a soft rag.
2. For heavy deposits, gently rub a fine-grained copper wool pad on the rollers until the polyethylene balls up and can be picked off by hand.

B. Heating shoes with nonstick surface (used in Laminex and Idex Machines):
1. Use a special polyethylene liquid remover available from your dealer. Wipe on with a soft rag.
2. Try fine-grained steel wool pad number 00 or finer. Rub very lightly, then blow away particles with compressed air (available at a camera dealer).

GLOSSARY

Core   Hard cardboard roll around which laminating film is wrapped.

Heating shoes   *See* Shoes.

Nip   Slit or space between the top and bottom laminating rollers where material is fed for lamination.

Off   Control used when lamination is finished.

PREHEAT or HEAT   Control used to warm up the machine.

Roller, rolls   See InDepth I, Major parts of the laminator.

RUN or DRIVE   Laminating controls. When RUN is pressed, the film will feed into the machine. Lamination can begin if the machine has reached proper temperature.

Shaft   Metal rod inside each supply roll that suspends the supply rolls on the machine.

Shoes   Panels that provide the heat for lamination on some machines.

Side plates   Circular metal plates on some machines that hold the shaft and supply rolls securely to each other.

Tacking iron   A small, hand-held accessory usually used in conjunction with the dry-mount press. When warm and applied to a heat-sensitive adhesive surface, it tacks or bonds this surface to another.

Threading knob   Found on some machines, used to carry sample cardboard through the machine before starting regular lamination with new rolls of film.

7 Thermal Transparency Makers

CONTENTS

Operating Tips	41	Suggested Settings	43
Problems	42	About Thermal Transparency Film	43
InDepth		Cleaning the Machine	44
Operating Procedure	42	**Glossary**	44

The thermal transparency maker is used primarily to make transparencies from paper originals for use on the overhead projector. Operation of the 3M Thermofax brand transparency maker is discussed here. You may find it useful to note that, in addition to transparencies, the 3M Thermofax also makes spirit masters for the ditto machine, individual paper duplicates and mimeographing stencils. It laminates, although results are not perfect.

OPERATING TIPS

1. Write, type, or print only with instruments that have carbon content, such as a lead pencil or carbon typewriter ribbon.

2. Run the original through a copier such as a Xerox machine before using the transparency maker, if the lettering or picture does not have enough carbon, is not dark enough, or is in color.

3. Unplug the cord when the machine is not in use, it is necessary to open the cover to clean the machine, or the machine does not cut off after the fan has run a fairly long time.

4. Remove paper clips, staples, and pins from your original.

5. Never insert cardboard or multisheet originals.

6. Use the carrier if the original is wrinkled, folded, or smaller than the copy sheet.

7. Write-on acetate film is ineffective in the transparency maker. Use only thermal film bought for this type of equipment (see InDepth III).

42   *Thermal Transparency Makers*

PROBLEMS

PROBLEM	CAUSE	SOLUTION
1. No print at all.	Original does not contain carbon.	Run original through a copier or use carbon typewriter ribbon or number 2 pencil.
	Transparency and original not sandwiched properly.	1) Hold transparency film so notch is in upper right-hand corner. 2) Place original face up, under transparency.
	Lamp burned out.	Replace lamp.
2. Black dots on transparency or paper copy.	Carrier belt, print roller dirty.	1) Clean (see InDepth IV). 2) If spots do not come out, replace belt.
3. Print emerges crooked on page.	Original not fed into machine evenly.	Line up original between two guide arrows on feed table.
	Film not laid straight in carrier.	Line up film in carrier with original.
	Print crooked on original.	Original must be redone.
4. Scorched carrier or original, or product torn.	Torn or damaged carrier belt.	Replace belt.
5. Material jammed in machine.	Improper feeding.	Lift machine cover and loosen carrier belt according to instructions on inside cover of machine. Remove jammed material, lock belt, and close cover.

IN-DEPTH

I. Operating procedure

A. Plug the cord into an AC outlet.

B. Insert any sheet of paper through the machine several times. This activates and warms up the equipment.

C. Hold the transparency film over the material so that the notched or rounded corner is in the upper right-hand corner.

D. Place the original image facing up underneath the film, or follow the instructions on the box of materials

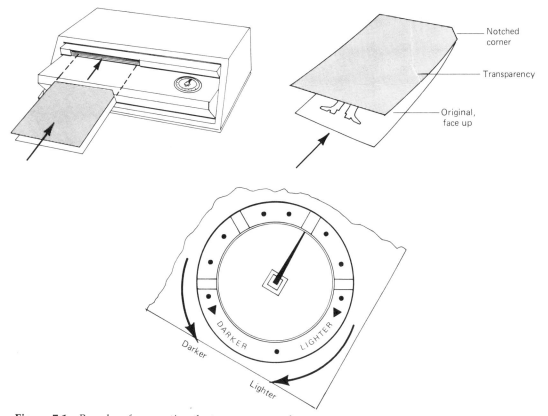

Figure 7-1 *Procedure for operating the transparency maker.*

you are using. Insert in the carrier if necessary.

E. Adjust the darkness dial (see In-Depth II).

F. Insert the material in the machine.

G. When finished, allow the fan to cool the machine.

H. Unplug the machine.

If the copy is too light or too dark, repeat steps C to F with fresh transparency film.

II. **Suggested settings for darkness dial**

Transparencies are on darker settings, while spirit masters are on lighter settings. Where available, colored markers on the dial help in dialing various functions. Generally, turning the dial counterclockwise produces a darker result; turning it clockwise produces a lighter result.

III. **About thermal transparency film**

Use thermal film put out by any company. Thicker film indicates better quality. Shelf life varies according to company.

Transparency film comes in two types: a sheet of clear film alone, or a sheet of clear film with tissue attached. Both produce the same results.

44 *Thermal Transparency Makers*

Films producing different color effects are available, as follows:

Clear film producing black print.

Clear film producing colored print (purple, green, and so on).

Colored film producing black print (film colors can be red, yellow, green, blue, pink).

Special needs film (consult your dealer).

IV. Cleaning the machine

Parts on the inside of the machine to be cleaned are the carrier belt and the shiny print roller.

1. Raise the cover.

2. Follow the directions for loosening the belt on the inside cover of the machine.

3. Using soft cloth, gently wipe the belt and roller with special Thermofax solution or alcohol. No rinsing is necessary.

4. Lock the belt and close the cover.

GLOSSARY

Carrier Plastic and paper envelope that encases materials before they are put through the machine. It keeps material together and helps protect it from damage.

8 Cameras

CONTENTS

Operating Tips	47
Camera Tips	47
Film Tips	48
Shooting Tips	48
Flash Tips	49
Problems	49
Cameras—Problems for 126, 110, disc, and 35 mm cameras	49
InDepth	57
General information—cleaning all cameras	57
Lens	57
Camera body	
Instant loading cameras: 126, 110, and disc formats	57
Steps in taking a picture	57
Loading film in the camera	58
Unloading film from the camera	58
Using flash	58
Films	59
35 mm cameras	59
Steps in taking a picture	59
Loading film	63
Rewinding and unloading film from the camera	64
Removing partially used film to use another type of film	65
Replacing partially used film in camera to finish unexposed film	65
Using flash	65
Film	70
Color film	71
Tinkering with exposure settings	72
Glossary	74

A camera is a light-tight box. The light-sensitive film you put in is held flat in the camera back. There is a lens in the camera front. The size of the lens opening and speed of the shutter regulate the light coming through the lens. The film is thus exposed to light and to what you want to photograph.

You can make your own very simple pinhole camera or spend hundreds on a complex one. An expensive camera can take poor pictures if not used correctly. The cameras covered here are the instant-load types (126, 110 cartridge, and disc formats) and 35 mm. Instant developing cameras are not covered.

Cameras vary in their features and in the adjustments you can make. In addition to picture composition, four factors interrelate and contribute to making every good picture.

1. Film sensitivity to light (ISO/ASA rating).
2. Size of the lens opening (aperture, also called f stop).
3. Speed at which the camera shutter admits and shuts out light (shutter speed).
4. Focus.

INSTANT-LOADING CAMERAS—126 AND 110 CARTRIDGE, DISC

The advantages of these cameras are low cost and ease of operation. Just point and shoot. On the basic models no exposure or focus adjustments are possible or necessary. Some of the more expensive cartridge and disc cameras incorporate features previously found only in 35 mm cameras, such as automatic exposure setting.

35 MM CAMERAS

Advantages over instant-load formats:

1. Many more types of film available for particular lighting conditions and special effects.
2. Superior lens quality for better picture detail. Interchangeable lenses available on many cameras.
3. Larger negatives produced. This provides larger photographs and more enlargement options.
4. Precise focusing.
5. Greater distance range when using flash.
6. Through-the-lens viewing of the subject on many cameras. What you see through the viewfinder is what you get rather than using a separate camera viewfinder.

Features. A wide range of 35 mm camera features is available, depending on the particular model.

Manually adjustable models. All settings, such as size of lens opening, shutter speed, and focus are made manually by the photographer. Manual settings are particularly useful because they provide the photographer with the flexibility to achieve special effects. A variety of lenses is available on single-lens reflex 35 mm cameras (SLR) in order to view and shoot subjects from different perspectives. Such lenses include various types of wide angle, telephoto, and zooms.

Automatic models. One or more of the settings is determined by the camera.

• Basic models—operate like instant-load format cameras. There is a fixed focus or slightly variable focus flexibility.
• Semi-automatic/manual models (A/M modes)—In semi-automatic (A) mode, the camera automatically sets precise aperture or shutter speed using electronic circuitry, and the user sets the other; in manual (M) mode, the user sets both aperture and shutter speed, according to built-in light meter settings if desired.
• Program/semi-automatic/manual models(P/A/M modes)—In program (P) mode, the camera automatically sets precise aperture and shutter speed and/or focus, using electronic circuitry; in the A and M modes, the camera operates like the semi-automatic/manually adjustable models described above.

Cameras that provide both automatic and manual kinds of options allow the user maximum convenience as well as photographic creativity.

In addition to automatic setting of exposure and/or focus, some cameras also have other automatic features: autoloading of film, autorewind when the film has been fully exposed, autoadvance from frame to frame, autoflash, and various audible signals or a voice synthesizer to remind you of your jobs!

Although automatic settings of various kinds work well under many conditions, they have limitations. For instance, with the program (P) mode that automatically sets the exposure, the camera cannot account for special lighting conditions such as a beach scene or snow. In this case the light meter will read the entire scene as very bright and program the camera to admit little light. That means your favorite aunt standing on the beach will look too dark or underexposed. You can override the camera's judgment using the manual (M) mode, reading the light meter and making the appropriate exposure adjustments yourself. On some cameras with the autofocus feature the camera will focus on the closest object reflecting light. Therefore, the cage bars, not the lion in the cage will be in focus. (Most autofocus cameras now can compensate for this problem. Some also include focus override for you to set focus manually.) With automatic film advance, you cannot remove a roll of film part-way, replace it with another type of film, and eventually finish the first roll. Nor can you create special effects with overlapping shots. Special conditions and special effects still require the photographer's knowledge, using manual settings.

OPERATING TIPS

CAMERA TIPS

1. Beware of careless handling, dust, moisture, heat, and sand.

2. Keep lenses clean.

3. Always cover the lens with a lens cap when not in use.

4. Whenever possible, keep the camera in a camera case, to protect camera and lens.

5. Make it a special point to use a camera case. If you forget to turn off your light meter, and light does not reach the meter's photo sensitive cell, there is little drain on the battery.

6. Do not keep the shutter cocked if the camera is not being used.

7. Occasionally operate the shutter if the camera is not used for long periods.

8. Do not let your fingers cover the lens or obstruct the electric eye, where available.

9. For 35mm cameras use an inexpensive sky-light filter or haze filter to protect the expensive camera lens from accidental blows and dirt.

10. Check the batteries, where available, to make sure they are good.

11. Clean the battery and camera contacts periodically.

12. When the camera is not in use, make sure the batteries are turned off where possible.

13. Remove batteries if the camera is not in use for one month or more.

14. Buy the correct battery for your camera. Camera manufacturers specify the particular size, type (such as silver oxide or mercury), and voltage batteries for their cameras. If these specifications are not met, camera functions, including the light meter settings, will not operate properly and may be damaged.

15. How exposure is set in the camera: On simple cartridge cameras, the exposure is keyed in when the cartridge is inserted. A light-sensing cell will give you a low-light

warning. If the film you have chosen cannot be interpreted by the keying system, you will have poor results. This happens if you use ASA 400 film not provided for on some 110 format cameras.

On more sophisticated cameras, a light-sensing cell determines the amount of light being reflected into the camera. You then either set the exposure yourself, or a computer determines and sets the exposure.

16. How focus is set—there are three basic ways to set focus.

Fixed focus—this is available on simple cameras. The focus is preset at the factory to provide optimum sharpness at normal distances, usually around 4 feet to infinity. The lens will not be as sharp close up or at infinity.

Manual focus—The user sets the focus by turning the focusing ring until the object appears in sharp focus in the viewfinder.

Autofocus—The camera sets the focus through an infrared sensing device. You may want a camera with a focus override so you can set the focus in special situations.

FILM TIPS

1. Do not leave film exposed to heat (such as in a car dash, glove compartment, or trunk).

2. Refrigerate film if it is not to be used for a long period. Let it stand for one hour before using.

3. Check the expiration date on the film box. Most film is good for six months after that date.

4. Have the film developed promptly.

5. Load and unload 35 mm film in dim light, not in sunlight. Do not worry about 126 and 110 cartridges or discs.

SHOOTING TIPS

1. Keep the camera from moving while shooting. Hold it steady. If necessary, brace yourself by leaning against a wall or placing one leg forward, one leg back, and pressing elbows against your chest.

2. Keep your subject from moving, unless your camera is equipped for action shooting. If your camera has adjustable shutter speed, always set it at 1/125 of a second or faster for action shots. Otherwise, the picture will be blurred.

You can freeze action (like a speeding train) through a higher shutter speed, or using a higher ASA film, or using an electronic flash on some cameras that allow a shutter speed of 1/125 second or more with flash. You then make the other appropriate corresponding camera settings.

3. Do not snap the shutter button. Squeeze down gently and release slowly.

4. Where your camera does not provide you with shutter speed or aperture flexibility, you can change the exposure by temporarily changing the ISO/ASA dial on your camera. This is to be used when you feel the camera's metering system cannot fully interpret the needs in a scene. For example, in a beach or snow scene, the camera will think little light should be admitted, but if there is a person in the scene the person will not be exposed with enough light. Here is the method:

a. The principle involved. ASA numbers are related to each other. ISO/ASA 200 film is twice as sensitive to light as ISO/ASA 100 film so it requires half the light. But this same ISO/ASA 200 film is half as sensitive to light as ISO/ASA 400 film so it requires double the light for good exposure.

b. Making the principle work.
   Determine the ISO/ASA number of the film you are using.

If you want more light, take half the number you are using (say you have ISO/ASA 200, cut in half is 100) and set it on your ISO/ASA camera dial. You have increased the light by one f stop.

If you want to compensate for a scene that is too bright, double the ISO/ASA number you are using (say ISO/ASA 200, doubled is 400) and set it on your dial. You have decreased the light by one f stop.

c. Reset your ISO/ASA dial to the original number after these special scenes have been shot.

FLASH TIPS

1. Test batteries for flash units at the camera store, drugstore, or hardware store.

2. Clean contacts on battery and flash unit with rough cloth or pencil eraser. Some deposits are invisible.

3. For units using batteries, there is a longer recycling time between flashes as well as reduced flash-subject range, if batteries are weak or terminals are dirty.

4. Flash recycling is affected by the type of batteries you are using. From fastest to slowest recycling are: nickel cadmium, alkaline and carbon zinc. *Do not use nickel cadmium where prohibited by the manufacturer. Damage will result to the flash unit.*

5. *Note*: The battery you are buying must be interchangeagble with what was recommended by the equipment manufacturer. Photo counters have battery replacement guides.

6. With the switch ON, and the ready light on, the battery is still discharging energy until the flash is fired. You are therefore using up battery energy. *But* when using NiCd batteries, do not discharge battery before storing, though the ready light is glowing. Turn ON/OFF switch to OFF.

7. If using flashbulbs, use lamp ejector after firing. The bulbs are very hot.

8. On inexpensive units, when the ready light glows, wait a little longer before shooting because on some flash units, the unit is not yet fully charged.

9. AC adaptors are made for the voltage requirements of a particular flash unit. They are not necessarily interchangeable. Check both before using.

PROBLEMS

PROBLEMS FOR 126, 110, DISC, AND 35 MM CAMERAS

PROBLEM	CAUSE	SOLUTION
While Shooting—Camera Problems		
1. Film will not advance.	Film already advanced.	Press shutter release.
	End of film.	Rewind 35 mm cassette. For 126, 110 cartridge, remove film when no paper is visible in film window. For disc, remove when "X" appears in film window.

50 *Cameras*

PROBLEM	CAUSE	SOLUTION
While Shooting—Camera Problems		
	On 35 mm autoload models: film improperly loaded.	Check 35 mm cassette that film leader is at designated autoload marker and that there is no extra slack in film leader.
		See also Problem 3, below.
2. Film will not rewind.	35 mm—rewind button not depressed before turning rewind crank.	Depress rewind button *before* rewinding film. It is usually on the bottom of camera.
	126, 110 cartridge and disc not made to rewind.	Remove 126, 110 film when no paper is in film window. Remove disc film when you see "X" in film window.
3. Shutter release will not work.	Film not fully advanced.	On cameras where you cock and release shutter manually, advance film lever as far as it will go.
	Motorized-advance models: film door and/or door lever not fully closed.	Close film door and door lever.
	On some models, built-in lens cover not fully open.	Open lens cover fully. See also Problem 1, above.
4. Light meter does not operate properly.	Meter not turned ON, where option available.	Turn meter ON.
	For semi-automatic (AUTO) and automatic cameras (PROGRAM), not following instructions.	Follow instructions in 35 mm InDepth I C or D (pages 62 and 63) so light meter works properly.
	Your type of meter, if CdS photocell, responds slowly to indoor/outdoor lighting change.	Wait a least two minutes before using meter.
	Long-life cell used up.	See repairman for replacement, if possible.
	Old batteries.	Replace batteries.

Cameras 51

	Batteries inserted incorrectly. Lens cover not opened.	Check battery diagram in camera. Open lens cover fully.
	Wrong batteries used.	Must use precise batteries specified by manufacturer according to type, size, and voltage for your camera.
5. Low light warning appears in viewfinder, where option available.	Need more light or more sensitive film.	Use flash or use film with higher ISO/ASA if available for your camera model.

While Shooting—Flash Problems

6. Electronic flash does not fire when camera shutter triggered.	Flash unit not ON.	Turn flash ON. Wait for ready light to come on.
	Ready light has not appeared where available.	Wait for ready light to come on. If it does not come on within a few seconds check other factors below.
	Weak batteries in flash unit.	Replace batteries.
	Where available, as in disc cameras, long-life cell used up.	See repairman.
	Battery contacts in flash unit and/or terminals on batteries need cleaning.	Clean with rough cloth or pencil eraser.
	Battery terminals reversed in flash unit.	Check battery diagram inside battery compartment.
	Separate flash unit not mounted on hot shoe, or sync cord not plugged into camera.	Attach cord from flash unit to camera where there is no hot shoe.
	Battery compartment on flash unit not closed.	Close compartment.
	Broken or improper flash cord, where cord is used. Defective flash unit. Defective camera.	Take in for repair.

52 *Cameras*

PROBLEM	CAUSE	SOLUTION
While Shooting—Flash Problems		
	Fully automatic flash units that come on when necessary sense sufficient light, and no need for flash.	Don't worry!
7. Electronic flash will not fire, after having just used it.	Unit needs normal recharging or recycling time.	Wait for ready light to come on.
	Unit not used for over a month. This means longer recharging time for the first few flashes.	Condition flash before shooting more pictures by letting it charge until ready, then fire. Repeat several times.
	If it takes over 30 seconds to recharge: weak batteries, dirty battery terminals, flash switch left ON during another occasion.	Replace batteries, clean terminals, turn flash switch OFF when flash not in use.
8. Flipflash, cube will not fire when shutter triggered.	Dead flash.	Flipflash: Turn flash to other half. Cube: Rotate one quarter turn.
	Defective flipflash, cube.	Try another flash cube of flipflash.
	Defective camera.	Take in for repair.
9. Red warning curtain in some instant-loading viewfinders for cameras that use cubes.	Cube already fired.	Rotate cube to another side; replace cube.
	Defective camera.	Take in for repair.
10. Flashbulbs do not fire when camera shutter triggered.	Dead flashbulb.	Replace bulb.
	Weak batteries in flash unit.	Replace batteries.
	Battery terminals reversed in flash.	Check battery diagram in battery compartment.
	Battery contacts in flash unit and/or battery terminals need cleaning.	Clean with rough cloth or pencil eraser.

Cameras   53

	Flash mounted on camera but no contact made—shoe not hot or sync cord not plugged in. (See Glossary).	Attach sync cord from flash unit to camera if no hot shoe on camera.
After Pictures Have Been Processed	*Determine Problem*	*Correct Problem, Shoot Picture Again*
11. No picture.	Lens cap on, or otherwise obstructed.	Remove lens cap. Keep hands, camera strap, case or other objects away from lens.
	Film did not pass through camera because the film was improperly loaded or there was no more unused film on the roll.	Check loading procedures. Use fresh film.
12. Pictures have incorrect colors.	Wrong film for lighting conditions.	See Instant Loading Cameras InDepth V (page 59); see 35 mm InDepth VII (page 70).
	Film outdated when shot or when developed.	Check expiration date on box before shooting film; develop promptly.
	Processing lab error.	Return negatives for reprinting. Do not need to reshoot.
	Film exposed to heat.	Do not leave film in or near heat source like car trunk, dash.
	Film exposed to radiation (for example at airport).	Purchase film shield to protect film.
13. Blurred pictures.	Camera moved (everything in picture blurred).	Brace camera against your body or use tripod; use faster shutter speed.
	Subject moved (subject-only blurred, rest is in focus. For example, your dog subject is blurred, but the tree background is clear).	Use faster shutter speed, if possible. Use 1/125 second or faster.
	Out of focus (some item in the scene is probably focused, but subject is out of focus).	Focus carefully on subject. Stay within camera focus zone.

54 *Cameras*

PROBLEM	CAUSE	SOLUTION
After Pictures Have Been Processed		
	Autofocus of inappropriate subject: dark hair, glary subjects, animal in cage.	Use prefocus lock on most autofocus cameras, or use manual override for special conditions.
	On two-position focusing system, used wrong position.	Select proper focus position: CLOSEUP or NORMAL.
	Subject closer than minimum distance at which equipment can focus.	All focusing systems: never shoot at closer range than minimum subject-to-lens distance specified for your unit. Note that with a fixed-focus camera, you cannot tell whether or not the picture is focused by looking through the camera.
		For 35 mm cameras that have manual focus, check feet scale on focus ring of lens.
	Picture-taking lens dirty. Blotches or specks may appear in picture.	Clean lens. (See InDepth, Cleaning All Cameras page 57).
14. Head or feet cut off.	Subject improperly framed in viewfinder.	Instant-loading and rangefinder cameras: Keep subject within luminous frame of viewfinder; back camera away from subject if necessary to fit subject into viewfinder.
		On two-position focusing system, set lens position for camera distance to subject.
15. Pictures too light/dark (poor exposure) *without flash*.	Poor lighting conditions.	Use flash.
	Electric eye obstructed.	Keep fingers off electric eye.

Cameras 55

	Your type of light meter (CdS Photo Cell) responds slowly to indoor/outdoor lighting change.	Wait at least two minutes to use CdS cell light meter. If hand-held meter, keep covered when not in use.
	Incorrect ISO/ASA camera setting.	Camera setting should match film box number.
	Camera warning for too much/too little light unheeded.	Too much light requires a change in shooting angle. Too little light requires flash or higher ISO/ASA type film.
	Wrong film for overall lighting conditions.	See Instant load InDepth V (page 59) or 35 mm InDepth VII (page 70).
	Light meter not read properly.	Learn how to use light meter.
	Light meter not working properly.	See Problem 4, above.
	Where applicable, f stop and/or shutter speed set improperly.	Take careful light meter readings. (Whether using manual or automatic camera make special f stop adjustments for special lighting as snow, beach, dark room.)
	For semi-automatic priority-setting cameras, did not use AUTO.	Use AUTO for f stop or shutter speed, whichever is appropriate for your camera.
	For program cameras, did not set and lock minimum aperture.	Set and lock minimum aperture.
16. Pictures too light/too dark (poor exposure) *with flash*.	Did not set ISO/ASA properly on flash unit.	Check film box; check setting on flash unit.
	Where applicable, f stop not properly set on camera.	See 35 mm, InDepth VI (page 65).
	Exceeded minimum and maximum distances for your flash unit.	For instant load cameras, see Instant load InDepth IV (page 58). For electronic flash, check owner's manual or dial on the back of the flash unit.

56 Cameras

PROBLEM	CAUSE	SOLUTION
After Pictures Have Been Processed		
	Electronic flash fired before unit was ready.	Wait for ready light or other indicator to come on.
	Some automatic electronic flash models: did not use range setting and switch indicated on flash unit.	For separately mounted electronic flash models, see 35 mm InDepth VI (page 65).
17. Only partial picture came out, using flash and 35 mm camera.	Camera shutter speed not on recommended flash settings.	Flash setting often indicated by another color on camera shutter speed dial. See owner's manual or try 1/60 second.
	Flash unit plugged into wrong camera outlet, where more than one outlet available.	See 35 mm InDepth VI C 1 (page 66).
18. Red eyes in color picture (white eyes in black/white picture).	Usually occurs under low light: 1. Flash mounted too close to lens. 2. Subject looking directly into camera lens.	1. Get extender for flash 2. Hand hold flash more than 4 inches from lens, and/or 3. Have subject look slightly away from lens, and/or 4. Point and shoot flash at ceiling if white, rather than at subject. Open lens 2 f stops.
19. Glare (using flash indoors or outdoors; without flash, indoors).	Reflection from glasses, mirror, or other shiny surfaces, with or without flash.	Move object, or shoot at angle, instead of directly at object causing glare.
20. Glare (without flash outdoors).	As above.	As above, or use polarizing filter. See photo dealer.
21. Color pictures have orange cast.	Daylight film used indoors under tungsten lighting.	Use flash with daylight film, or use tungsten film.
22. Color pictures have bluish cast.	Tungsten film used outdoors.	Use daylight film.

IN-DEPTH

GENERAL INFORMATION—CLEANING ALL CAMERAS

I. Lens

Remove lens, where possible, and clean lens at either end. Note that lens coatings produced within the last five years don't damage as easily as previous coatings.

A. Dust off with a can of compressed air, then use a camel's hair brush.

B. Breathe on the lens to create a mist.

C. Clean with lens tissue, *not* eyeglass or facial tissue. Fold into a soft pad with no wrinkles, and wipe gently.

D. Repeat step B. If mist clings to an area there is an oily spot such as a fingerprint. Use a bottle of commercial lens cleaner. Put a drop of fluid on lens tissue, *not on the lens*. Use lens cleaner as rarely as possible.

II. Camera body

A. Outside, wipe the body with clean, soft, dry cloth.

B. Inside, open the back cover. Never touch the shutter curtain. Use compressed air and a camel's hair brush for other areas. When using compressed air, beware of moisture that has accumulated in the can. Test the air on your hand. Hold can about 6 inches away. When lens has been removed: Never touch mirrors. Let an expert clean this area.

INSTANT-LOADING CAMERAS: 126, 110, AND DISC FORMATS

I. Steps in taking a picture

A. Decide which film you want to use.

B. Load the film.

C. Remove the lens cap, where applicable. Turn camera ON, if necessary.

D. Determine whether the light situation requires flash. On some instant-loading cameras, slightly depressing the shutter release will activate a light in the viewfinder to tell you to use flash. On other units, such as many disc cameras, the unit will automatically activate the built-in flash when necessary.

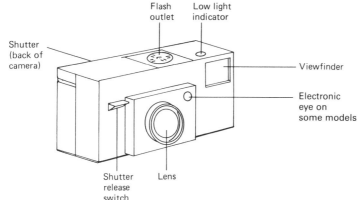

Figure 8-1 *Basic elements of an instant load camera.*

E. Compose the picture so that it is within the luminous frame of the viewfinder.
F. Advance the film lever until it locks.
G. Manual focusing is often not possible. Switch to CLOSEUP where applicable.
H. Point and shoot. For many instant loading cameras, shoot between four feet and infinity.
I. If the film is removed from the camera before the last shot in the 126 or 110 cartridge has been taken, you may lose the exposure you last photographed.

II. Loading film in the camera

A. Open the back cover of the camera.
B. Insert the cartridge or disc. It will only fit one way.
C. Close the cover.
D. Manual advance cameras. Flip advance lever as far as it will go; allow it to return. Repeat until the lever locks or until you reach number "1". Thereafter, operate film advance lever to the next exposure until it locks.
E. Motorized advance cameras (disc and some cartridge models). After closing the film cover and opening the lens cover, some cameras automatically advance the film to the first frame and to subsequent frames.

III. Unloading film from the camera

A. For 126 and 110 film use the film until all the paper is gone from the film window. For disc film, an "X" will appear in the film window at the end of the film.
B. Open the film cover of the camera.
C. Remove the cartridge or disc but do not tamper with it.
D. Close the film cover of the camera.

IV. Using flash

A. Types of flash (see top table, page 59)
   1. The instant-load model determines whether to use regular cubes, high power cubes, flip-flash, or electronic flash. Many cartridge-load models and all disc cameras have built in electronic flash units. Their flash capabilities vary widely. Know your model.
      a. Automatic flash? Does your camera advise you when to turn on the flash or does it activate the flash automatically according to low light conditions?
      b. Instant recycle or recharge? Can you shoot a second flash picture after just having used the flash, or must you wait until the camera tells you the flash is again ready to be used? Most disc cameras have almost-instant recycle.
      c. Batteries? Does your camera need batteries to operate the electronic flash or does it have long life cells to be used for many years? Many disc cameras have long life cells.
B. Subject-flash distance
   1. The most important factor affecting proper flash exposure is the distance from subject to flash.

Regular Cubes	*Magicubes*	*Flipflash*
Four flashes per cube	Four flashes per cube	Eight or ten flashes
Camera itself rotates cube when camera lever is advanced to next exposure.	Camera itself rotates cube when camera lever is advanced to next exposure	After one side has been used, remove flipflash. Flip it so fresh bulbs are on top.
Check camera battery every three months.	No flash battery necessary in camera.	No flash battery necessary in camera.
If one dead flash, rotate cube by hand.	If one dead flash, rotate cube by hand.	If one dead flash, flip flipflash to other half.

Suggested camera-subject distances with flash		
Regular Cubes	Minimum 4 Feet	Maximum 9 Feet
High Power cubes	6	15
Flipflash	Check with your camera manual or dealer	
Disc camera with built-in electronic flash (closeup lens position)	4 18"	18 4"
Other instant-load electronic flash	Check with you camera manual or dealer	

V. **Films** (see table, page 60)

   A. How to tell print from slide film. The term COLOR in the film name = print-type film, CHROME = slide-type film. Thus, Kodacolor is print film, while Kodachrome is slide film. One exception is Verichrome Pan, which is print film. For an extra charge you can convert slides into prints, and the negatives of prints into slides.

   B. What films are commonly available. Cartridge and disc film can be used both indoors with flash, and outdoors. 126 and 110 print films comes in 12 and 24 exposures, and slide film comes in 20 exposures. Disc comes in 15 exposures.

35MM CAMERAS (see Figure 8-2)

   I. **Steps in taking a picture**

   Note: Determine whether your camera is automatic, adjustable or has both options. If you are not sure, check the Introduction of this chapter for a description of each camera type. Use the appropriate section below.

   - Adjustable cameras (See A below.)
   - Basic automatic cameras (See B below.)
   - Semi-automatic/manual cameras (See C below.)
   - Program/semi-automatic/manual cameras (See D below.)

   A. Adjustable cameras

      1. Before following the procedures below, find out where to set the ISO/ASA, f stop, and shutter speed on your camera. Learn how to use a light meter. (See also Figure 8–3 and InDepth Chapter IX.)
      2. At a camera store or drugstore, check the light meter batteries of the camera where available.
      3. Decide on the appropriate

Films for Instant Loading Cameras

	Print (B&W/Color)	ISO/ASA*	Slide (Color)	ISO/ASA*
126	B&W—Verichrome Pan	125	Kodachrome 64**	64
	Color—Kodacolor II	100	Ektachrome 64**	64
	Kodacolor VR	200	Ektachrome 200	200
	Kodacolor 400	400		
110	B&W—Verichrome Pan	125	Kodachrome 64	
	Color—Kodacolor II	100 (discontinued)		
	Kodacolor VR	200		
	Kodacolor 400	400		
Disc Kodacolor VR (15 exposures)		200	None	

*ISO/ASA indicates film's sensitivity to light (See Glossary).
**Kodachrome 64 and Ektachrome 64 can be used interchangeably. However, Kodachrome often takes more time to process than Ektachrome because it is sent away to Kodak laboratories, while Ektachrome can be processed locally. Also, Kodachrome emphasizes warmer tones, Ektachrome cooler colors.

film to use: black and white or color, prints or slides, daylight or indoors, film sensitivity to light (ISO/ASA). (See InDepth VII.)

4. Adjust the ISO/ASA setting on the camera. See the film box for the ISO/ASA of the film.

5. Load the film.
6. Turn the camera light meter on where possible.
7. Remove the lens cap and/or shield.
8. Take a light reading with the built-in light meter or a hand-held meter.
9. Set the f stop and shutter

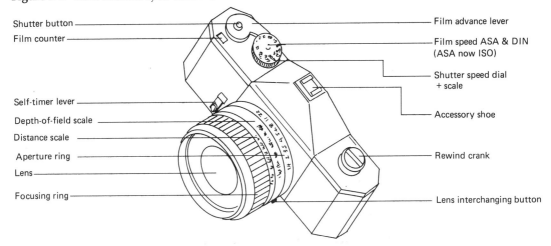

Figure 8-2 *Basic elements of an instant 35mm camera.*

Shutter button
Film counter
Self-timer lever
Depth-of-field scale
Distance scale
Aperture ring
Lens
Focusing ring

Film advance lever
Film speed ASA & DIN (ASA now ISO)
Shutter speed dial + scale
Accessory shoe
Rewind crank
Lens interchanging button

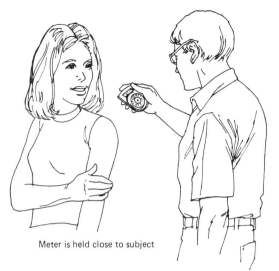

Meter is held close to subject

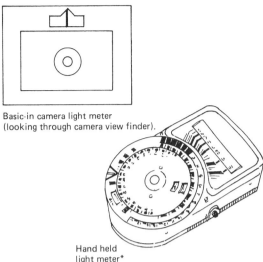

Basic-in camera light meter (looking through camera view finder).

Hand held light meter*

Figure 8-3 *Using a light meter.* For built-in meter: (1) Set ISO/ASA on camera, turn meter* ON, *if necessary. (2) Hold camera close to subject, usually from the same direction you will shoot picture.** (3) On non-automatic cameras, adjust f stop ring and shutter speed dial until viewfinder meter needle is centered. If you prefer a certain f stop or shutter speed, set it first. For comments on exposure, see also Glossary: Aperture, Shutter Speed, Depth of Field. For hand-held meter: (1) Set ISO/ASA on light meter. (2) Hold meter close to subject from the same direction you will shoot picture.** (3) Rotate marker on meter to where needle points. (4) There will be several correct f stop/shutter speed exposures shown. Make your selection.*

(*The meters shown are the popular reflected light meters. Incident light meters are not covered in this book.   **Meters read all tones, whether dark, light or gray tones. Usually you want the proper original contrast for your picture. You may need to meter both the light side and the dark side of a subject if they are substantially different, and take an average of the two. If metering a very dark subject, set camera one f stop smaller than meter indicates. If metering a very light subject like snow, open f stop on camera by 1-2 f stops more than meter indicates.)

speed according to the light reading, or make necessary compensations for special lighting. Where low light is indicated by the light meter and the subject is not moving, mount the camera on a tripod.
10. Compose the picture.
11. Focus. Turn the focusing ring while looking through the viewfinder until the image is clear in the focus zone.
12. Advance the film lever, release, and then press the shutter release.*

*If the shutter release does not work, advance the film lever as far as it will go. If the advance lever does not advance at all, press the shutter release first.

13. Repeat steps 8 through 12 for each picture.
14. When finished shooting, turn the light meter off where possible.
15. Replace the lens cap.
16. Push the rewind button and rewind the film.

B. Basic automatic cameras

1. Decide on appropriate film to use: black and white/color, print/slides, daylight/indoor, film sensitivity to light. (See InDepth VII.)
2. Remove lens cap and/or shield where possible. Where avail-

able turn main switch ON; activate light meter.
3. Set ISO/ASA on camera. See film box for ISO/ASA of your film.
4. Use camera battery check, where available, to test functioning of light meter or electric eye. Otherwise, periodically check batteries at camera store. Weak batteries will result in inaccurate light readings and therefore poor exposure of your pictures.
5. Load film (see InDepth II.)
6. Compose pictures within luminous frame or inset markers of viewfinder.
7. Focus. Where available, turn focusing ring while looking through viewfinder until image is clear.
8. Advance film lever, release.
9. Take steps appropriate for your camera to determine if there is enough available light for a good exposure. With some cameras insufficient light is indicated when you lightly depress the shutter button. If you need more light, use either flash (see InDepth VI) or film with higher ISO/ASA number, such as ISO/ASA 400 instead of 100.
10. Shoot your picture by depressing shutter button.
11. Repeat steps 6–10 for each picture.
12. When finished replace lens cap.
13. Push rewind button and rewind film.

C. Semi-automatic/manual cameras

Semi-automatic mode

1. Follow InDepth, Chapter I B, Steps 1–7.
2. Set aperture or shutter speed, depending on which is set manually for your camera model.
   a. Camera with shutter-speed priority. This kind of camera is particularly useful when you take many action shots.
      —Set desired shutter speed manually.
      —Set aperture (f stop) on A (AUTO) or EE (electric eye). The aperture is then automatically coordinated with the proper shutter speed for correct exposure.
   b. Camera with aperture-priority. This kind of camera is particularly useful if you like to create moods in your picture. (See Glossary, Depth of Field, Aperture.)
      —Set desired aperture (f stop) manually.
      —Set shutter speed dial to AUTO. Shutter speed is then automatically coordinated with the proper aperture for correct exposure.

In both semi-automatic camera types you can check aperture/shutter speed combination in the viewfinder. As lighting changes occur, the camera will continuously adjust the AUTO setting to coordinate with your manual setting. If you change the manual setting, the AUTO setting will change accordingly.

3. Follow steps 8–13, InDepth I B.

Manual mode

Disengage aperture ring or shutter speed from A (AUTO) setting. Follow direction in InDepth I A.

D. Program/semi-automatic/manual cameras

Program or fully automatic mode

1. Follow InDepth I B, Steps 1–5.
2. Set camera on P mode.
3. Prepare aperture ring. On some cameras set aperture ring at A; on some cameras set aperture ring at minimum lens opening (usually f/16 or f/22). Aperture ring should be locked with lock button.
4. Compose picture within luminous frame or inset markers of viewfinder.
5. Focus.
6. Advance film lever, release. Depress shutter release to shoot your picture.
7. When film is finished, push rewind button, rewind film, turn camera OFF, replace lens cap.

Semi-automatic mode

1. Disengage P mode.
2. See InDepth I C.

Manual mode

Disengage Program (P) or Automatic (A) modes.
Follow steps in InDepth I A.

II. **Loading film**

Autoload of film. Some cameras have automatic loading of the film onto the take-up spool. Follow manual loading directions A–E below. Then place film leader at designated autoload marker, close camera back, press shutter release up to the first shot. You must carefully follow directions for proper placement of film leader. Improper placement of film leader or extra slack in film can cause autoload malfunction.

Manual loading of film. Load film in the shade. Light leaks may result if you don't.

A. Remove the camera case, if there is one. It often unscrews from the bottom.
B. Open the back cover (how it opens depends on the individual camera). Try pulling up slightly on the rewind knob.
C. Hold the film cartridge with the shiny side of the film leader facing up.
D. With the camera face down, place the film cartridge into the slot on the left side of the camera (in some cases, adjust—push or pull—the rewind crank on the top left outside the camera before inserting the cartridge, so that the film cartridge can drop into the slot).
E. Locate the take-up spool inside the camera. This spool has one or more slots. See Figure 8–4a.

64 Cameras

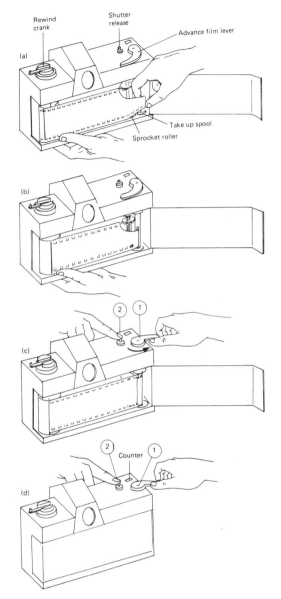

Figure 8-4 *Loading your camera.*

F. Also locate the sprocket roller, near the take-up spool. It has teeth (sprockets) at each end.

G. Feed the film leader into one slot of the take-up spool. See Figure 8–4b.

H. Keeping the back open and the camera face down, advance the film lever all the way, release, and then press the shutter release.* Do this two times until the holes on both sides of the film are snugly into the teeth of the sprocket roller. See Figure 8–4c.

I. Replace the back cover of the camera and close. Locate the film counter, usually at the top right of the camera.

J. Advance the film lever, release, and then press the shutter release twice. See Figure 8–4d.

K. You have now properly loaded your film.
(Usually, the film is advancing properly if you can see the rewind crank turning by itself counterclockwise while the film lever is being advanced.)

L. Before shooting your picture, refer to the steps in taking a picture, above, for either automatic or adjustable cameras.

III. Rewinding and unloading film from the camera

A. Lift the lever on the rewind crank.

B. Activate the rewind release (usually by pushing in the small button on the bottom of the camera).

C. Turn the rewind crank, usually clockwise. This will rewind the film back into the cartridge.

D. As you turn the crank, there is a slight resistance, indicating that the film is leaving the take-up

*If the shutter release does not work, advance the film lever as far as it will go. If the advance lever does not advance at all, press the shutter release first.

spool. Rewind until you feel no tension at all on the rewind crank.

E. Open the camera back. Remove the cartridge. Close the camera.

IV. **Removing partially used film to use another type of film**

(Special room conditions are not necessary.)

A. Note the number of the last exposure on the camera film counter.

B. Unfold the rewind crank.

C. Activate the rewind release (usually by pushing in the button on the bottom of the camera).

D. Turn the rewind crank, usually clockwise, only until slight film resistance is relieved in winding. Do not rewind further; otherwise, the film leader will go back into the film cartridge. If the leader goes back into the cartridge, buy a film leader retriever from your camera dealer for a couple of dollars. The leader is necessary to reload the camera.

E. Remove the cartridge from the camera.

F. Write the film counter number on the film leader of the cartridge, for your reference.

G. Store the cartridge, preferably in the refrigerator.

V. **Replacing partially used film in camera to finish unexposed film**

(Special room conditions are not necessary.)

A. Before placing the film cartridge in the camera, note the number of the last exposed shot marked on the film leader.

B. Allow the film to warm up if it was stored in the refrigerator.

C. Keep the lens cap on the camera lens. Load film.

D. Advance the film lever, release, and press the shutter. Repeat several times, up to the last exposed shot you noted on the film leader. Read the film indicator at the top of the camera.

E. Advance the film lever and press the shutter release for two more shots, so that there is no chance of double exposure on the original shots.

F. You have now properly reloaded your unfinished film.

G. Remove the lens cap and resume shooting.

H. Before shooting your next picture, follow Steps in Taking a Picture, above.

VI. **Using flash**

A. General information
The main purpose of using flash is to increase light on the subject being photographed when existing light conditions are not sufficient. Every flash unit has a minimum and maximum distance within which it is effective. Never exceed the limits for your unit. Otherwise, you will obtain pictures that are too light or too dark. Depending on your 35 mm camera, use built-in flash or separate flash unit mounted on the camera.

B. Built-in flash units

Most units are simple to operate. A few cameras now have totally automatic flash like many instant-load disc cameras. As you depress the shutter release, the camera will use flash if necessary. In most cases, the camera alerts you when there is too little light, usually by a signal in the viewfinder. Turn the flash switch ON. When the ready light appears in the viewfinder, shoot your picture. Exposure is automatically set for the specified subject-to-camera distances of your unit. When the ready light appears again, take your next picture.

C. Separate flash units

There are three types of separate flash units: dedicated electronic flash, conventional electronic flash and flashbulb units. Cameras differ widely as to the settings and connections required for flash use. General guidelines are given below. You may need to consult your owner's manual for more details.

1. Connecting flash unit to the camera. When you attach the flash unit to the camera, synchronize it with the camera's shutter release system.
    a. Hot shoe connection. Mount flash unit onto hot shoe where it is available (See Glossary, Hot Shoe). If there is no hot shoe, use the sync cord method.
    b. Sync cord connection. Mount flash unit onto plain shoe, or on bracket which you attach to cam-

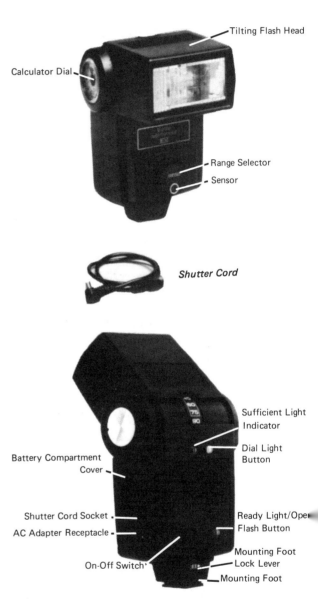

Figure 8-5 *Front and back view of a conventional electronic flash. (Courtesy Vivitar Corporation)*

Cameras **67**

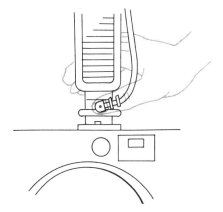

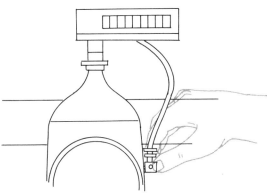

Figure 8-6 *(a) With hot shoe on camera. Place flash cord on flash pin provided, if cord is permanently attached to flash. (b) Without hot shoe on flash. Attach flash cord to camera outlet provided.*

era. Connect flash unit cord to the appropriate camera sync outlet. Sometimes the camera outlet is covered with a rubber tip that must be removed.

Selecting the appropriate camera sync outlet or switch for flash: Most cameras have sync outlets for both flashbulbs and electronic flash. This is because not all flash types reach peak light output immediately after the flash is fired. The particular outlet or switch controls flash firing and opening of the camera shutter, so that the film receives peak light output when the shutter is fully opened. Common outlet/switch designations are X—electronic flash; M, F, FP—various types of flashbulb sync settings. Ask your dealer to check the flashbulb box to see that the bulbs are appropriate for your camera. (M is for leaf type shutters; F, FP are for focal plane shutters).

2. Setting camera/flash
   a. Using dedicated electronic flash units (see Glossary, Dedicated Flash).
      —Set ISO/ASA on camera and on flash unit where required.
      —Set camera/flash mode, where available, and depending on your camera. Use Program (P) mode for automatic camera/flash settings of aperture and shutter speed. Use Automatic (A) if you determine one of the exposure settings and the camera/flash determines the other. Use Manual (M) if you are to set both exposure set-

tings yourself. See conventional electronic flash, below.

b. Using conventional electronic flash or flash bulbs:
—Set ISO/ASA on camera and on flash unit.
—Set camera shutter speed. A usual setting for electronic flash is 1/60 second marked in a different color on the camera shutter speed dial. Check specifications for your camera model. For flashbulbs, an average shutter speed is 1/30 second. However, check flashbulb box for the particular ISO/ASA film you are using.
—Determine flash-subject distance.
—Set proper camera aperture using f stop or guide numbers, depending on your camera, as explained below.

c. Determining camera aperture for conventional flash or flash bulbs.
—Determine camera f stop (where available on camera). See Figure 8–7. Sample numbers on camera: 1.4, 2, 2.8, 4, 5.6, 8, 11, 16.

Figure 8-7 *Electronic flash dial for determining f stop on camera.*

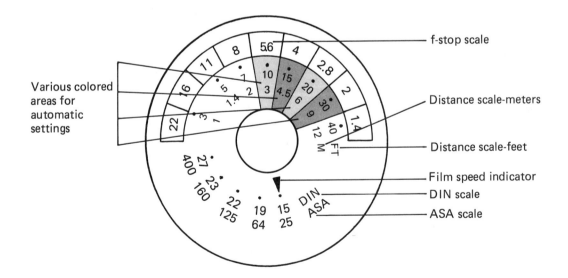

Electronic flash, manual setting: (1) Put flash unit setting to M., (2) use the chart or dial on the back of the flash unit, (3) locate the ISO/ASA of your film on the flash unit chart or set the ISO/ASA on the flash dial; (4) measure the distance from flash to subject in feet. Locate it on the chart, or dial it; (5) the chart or dial will then indicate the f stop; (6) set the f stop on the camera.

Electronic flash, automatic setting (recognized by color-coded exposure dial): (1) Set the ASA on the flash dial; (2) measure the maximum flash-to-subject distance in feet; (3) on the flash dial, look for the color that matches your distance range; (4) the colored area will indicate the f stop; (5) set the f stop on the camera; (6) set the flash switch to the color that matches the color selected on the dial.

Flashbulbs. (1) Determine the film ISO/ASA; (2) determine the guide number for your reflector and bulb type from the film sheet or flashbulb package; (3) measure the maximum flash-to-subject distance in feet; (4) the guide number divided by the distance equals the f stop. Example: guide number 80 ÷ 10 feet = f8; (5) set the f stop on the camera.

—Determining camera guide numbers (GN) where there is no camera f stop.

Sample guide numbers on camera: 22, 32, 45, 65, 90, 130, 180.

Use the electronic flash on the manual setting, where there is automatic/manual switch.

With electronic flash unit dial: (1) Set the ISO/ASA on the flash dial; (2) locate the 10-foot mark on the dial and its corresponding f stop; (3) multiply that f stop by 10. Example: f5.6 × 10 = guide number 56; (4) set the guide number (GN) on the camera.

Without electronic flash unit dial and chart: (1) Talk to your camera dealer, or write to the flash unit manufacturer to get the BCPS (beam candlepower seconds) of that unit; (2) use the film sheet that comes with a particular roll of film to find the guide number for your BCPS; (3) set the guide number on the camera; (4) once you have the BCPS, you will be able to find the guide number for any film you

are using with that flash unit.
3. Operating camera/flash unit
   a. For all electronic flash units unless using AUTOflash with sensor that decides when to flash.
      —Turn unit ON where possible.
      —Wait for ready light to appear on flash unit. In some cases, light will also appear in camera viewfinder.
      —Advance film lever, release. Shoot picture.
      —Wait for ready light to reappear. This indicates that the flash has recycled.
      —Advance film lever, release. Shoot another picture.
   b. For all flashbulb units:
      —Insert flashbulb.
      —Shoot picture.
      —Remove bulb carefully, holding head of bulb with tissue if bulb is hot. Insert new bulb. Shoot another picture.

VII. Film

A. Selecting the correct film for your needs.
   1. The 35 mm format produces a rectangular picture. There is a large selection of films for both general and special applications. Since every film has special exposure and print characteristics, some experimentation is advisable when you are just beginning. You can eventually select a few types of film and use them consistently.
   2. The 35 mm format produces a 3-1/2 inch × 5 inch photo which can be enlarged to 4 × 6, 5 × 7, 8 × 10, 8 × 12, 11 × 14, 16 × 20.

   When selecting film, consider these factors:
   3. If you want COLOR FILM, the films are commonly available in:
      a. Prints (daylight-type—usually use flash indoors). The term COLOR is usually in the name, as in Kodacolor.
      b. Slides (daylight or tungsten indoor). The term CHROME is usually in the name, as in Kodachrome.
      c. Various ISO/ASA ratings.
      d. Many shots or few shots: 36, 24, 20 exposures, depending on the film type.
   4. If you want BLACK AND WHITE FILM, the films are commonly available in:
      a. Prints only, daylight type only. The film can be used outdoors or indoors with/without flash depending on film sensitivity and lighting conditions.
      b. Various ISO/ASA ratings.
      c. Many shots or few shots: 36, 24, 20 exposures, depending on the film type.

B. Notes on prints/slides; ISO/ASA ratings:

1. Prints or slides. The photo lab can make prints from slides. Slides can be made from print negatives or prints themselves.
2. Selecting film with the correct ISO/ASA can be difficult. Listed below are some suggestions.

General Guidelines for Selecting ISO/ASA Film Speeds

The higher the film speed, the more sensitive it is to light. The slower the film speed, the sharper the picture produced.

Available conditions or desired results. (Use appropriate camera settings. See also InDepth VIII.)	*ISO/ASA Speeds* (See also other properties of particular films.)
Freeze action (A speeding train will not look blurry)	*Very High Speeds* ASA 400, 1000
Only low light available Good depth of field possible	*High Speeds* ASA 160, 200, 320
Average outdoor conditions, or indoor conditions with flash Good enlargements	*Medium Speeds* ASA 64 to 125
Bright lighting Good depth of field Best enlargements	*Slow Speeds* ASA 25, 32

VIII. Color film

A. General applications
1. Daylight film—can be used outdoors, indoors with electronic flash or blue bulbs, indoors without flash if 75 percent of existing light is daylight.
2. Tungsten indoor film is used with little daylight and mostly artificial lights, such as lamps. You may use clear, not blue flash bulbs. Do not use electronic flash.

B. Color film with electronic flash units. When using flash, always use daylight film.

C. Color film for fluorescent lighting. Use film marked daylight. The resulting greenish hue is more acceptable than the very blue pictures produced with tungsten film. If not using flash, use an FLD filter. When using flash do not use a filter.

D. Color film with flood lamps. Buy your favorite color film, then ask your camera dealer for the appropriate flood lamps. Flood lamps to provide extra indoor lighting for studio or copy stand use come in daylight and two kinds of tungsten. Your lamps should always match the kind of film you are using. (See Glossary, Kelvin, Color Balance.) For quick reference, note that tungsten bulbs are clear and daylight bulbs are blue.

E. Filters for color films. Light has color. The colors of daylight, household lamps, studio lighting and fluorescent light are all different. When films are marked daylight or tungsten, they are accounting for some of the color variations under different lighting conditions.

Adapting Color Film to Varying Lighting Conditions.

FILM BOUGHT IS:	PICTURE TO BE TAKEN UNDER:	ADD FILTER:	OPEN LENS, IF YOU DO NOT HAVE BUILT-IN METER ON CAMERA:
daylight film	tungsten lights, or no flash	80 A	two f stops
	household lamps, other nondaylight tungsten lighting, with flash	no filter	set flash and camera
	fluorescent lights, no flash	FLD or 30 magenta	one f stop one f stop
	fluorescent lights, flash	no filter	set flash and camera
tungsten film	daylight	85 B	one f stop

Filters are available to allow the use of the film under different lighting conditions from those for which it was intended. The filters screw onto the front of the camera lens. Because the use of filters changes the density of light reaching the film in the camera, the aperture or lens opening of the camera must allow more light in. A built-in light meter will make necessary adjustments in the lens opening, once the filter has been attached to the lens.

IX. **Tinkering with exposure settings**

A. Aperture and shutter speed are the two exposure settings. You can determine proper exposure for a scene using a built-in or hand-held light meter. On occasion you may need to compensate for light meter readings if there are special conditions, as explained below.

B. The aperture has two jobs. The first is to provide proper exposure by controlling *the amount* of light that hits the film. The second is as a major factor in determining depth of field. Both jobs occur simultaneously. Aperture settings control the size of the lens opening.

C. The shutter speed also has two jobs. One is to provide proper exposure by controlling *how long* a given amount of light hits the film. The other is to freeze (or blur) action of a moving object. Both jobs occur simultaneously. Shutter speed settings control how long the camera shutter stays open.

D. Aperture and shutter speed are directly related to each other and work together to control the light admitted through the lens to hit the film.

E. Here is how it works. Assuming you have set the ISO/ASA, every change from one shutter speed to the next will double or cut in half the light reaching the film. Also, every change from one aperture setting or f stop to the next will either double or cut in half the light reaching the film.

How Aperture and Shutter Speed Work
SIMULTANEOUS JOBS:
PROPER EXPOSURE—JOB 1
CREATIVE EFFECT—JOB 2

Aperture

Halves incoming light—Job 1 *Greater depth of field—Job 2* With each change to the left from f 2 to f 16 [Lens aperture is made smaller to admit half as much light; "closing" or "stopping down" lens]	*Doubles incoming light—Job 1* *Less depth of field—Job 2* With each change to the right from f 16 to f 2 [Lens aperture is made larger to let in twice as much light; "opening" lens]

Sample light meter reading (after ISO set)

(f stop)						
16	11	8	**5.6**	4	2.8	2
(Fraction of a second)						
15	30	60	**125**	250	500	1000

Shutter Speed

Doubles incoming light—Job 1 *Moving objects more blurred—Job 2* With each change to the left from 1000 to 15 [Shutter opened twice as long; "slower speed"]	*Halves incoming light—Job 1* *Moving objects less blurred—Job 2* With each change to the right from 15 to 1000 [Shutter opened half as long; "faster speed"]

F. If you need to make a change in exposure, move one setting, not the other. Do this if you think that for special reasons there should be more or less light hitting the film than the light meter indicates. For example, if you are shooting a scene at f 5.6 at 1/125 of a second, changing the f stop to f 8 at 1/125th of a second will change the exposure. The aperture will be smaller and less light will reach the film.

G. If you want to make a creative change *but* keep the exposure the same, you must change both settings correspondingly, so that the same amount of light as originally determined by the light meter will reach the film.

1. For depth of field: change aperture setting first, then set corresponding shutter speed.
   a. In our example, if f 5.6 at 1/125 of a second is changed to f8 for greater depth of field (more of the total picture in focus) make a corresponding shutter speed change to 1/60 of a second. Since less light is coming through the aperture, more light must be allowed through a slower shutter speed.
2. For freezing an object in motion: change the shutter speed

first, then set the corresponding aperture setting.

a. In our example, if the shutter speed is increased from 1/125 of a second to 1/500, change the f stop from 5.6 to f 2.8. Less light is allowed in through the shutter, more light must be allowed in through a wider aperture. (While your main goal was to freeze action, you lessened depth of field because aperture and shutter speed coordinate.)

This means that f 5.6 at 1/125, f 2.8 at 1/500 and f 8 at 1/60 all will give you a properly exposed picture in our example, but the creative effects will be different.

GLOSSARY

ASA A number on film box indicating the film's sensitivity to light. Now ISO is used. (See ISO.)

Aperture Also called iris, opening, diaphragm. Lens opening. Size of opening can be controlled on many cameras except for most instant loads. Adjusting lens opening affects:

1. The amount of light reaching the film, and
2. The depth of field in the picture; that is, what will be in focus in the picture of a given scene.

Sizes of lens openings are measured in f stops. (See Glossary, f stop.) Aperture settings are coordinated with shutter speed settings.

Available light Also called existing light. Normal illumination that photographer finds existing where photos are to be made. The photographer uses no auxiliary lighting, as flash or photofloods. Lighting conditions in the following would be available light:

Daylight, outdoors

Indoors at night in homes, as table lamps

Public buildings

Stage shows

Church interiors

Outdoors at twilight

After dark on lighted streets

Daylight indoors with added artificial light, as table lamps

Christmas lights

Candlelight

BCPS (Beam candlepower seconds) A unit of measure for the output of light by an electronic flash unit. Provided in owner's flash manual (See InDepth VI, page 69).

Color balance Refers to the color emphasis of light (blue or red) to use with the appropriate film when shooting color prints or slides. (See also Glossary, Kelvin.)

1. For daylight conditions, use daylight film.
2. For electronic flash, use daylight film.
3. For fluorescent lighting, use daylight film with FLD or 30 magenta filter.
4. For household lamps, use tungsten film with photofloods or clear flashbulbs.

Compressed air Can of air available at a photo dealer. Used for cleaning delicate parts of the camera or lens that should not be touched too often, if at all. Do not shake;

hold upright when spraying. Test on your hand before using on camera.

Contact print An 8 × 10 print composed of small individual shots of an entire roll of film in the sequence in which they were shot. You may select best exposures for regular size prints to be made. Contact prints can be ordered from your camera store upon request for a small charge.

Dedicated flash A type of separate add-on, not built-in flash unit that integrates flash photography with camera functions. A particular flash model is "dedicated" to a brand and model series of camera to integrate with its electronic circuitry. For example, a Focal 200 flash unit is dedicated to Cannon A series cameras.

When used with dedicated flash cameras, the flash unit uses the camera's electronic circuitry to automatically set the camera's shutter speed and/or aperture. On many cameras, the flash-ready signal shows on the camera viewfinder's LED, in addition to the lamp on the flash unit. The camera's photocell allows just enough flash light for proper subject exposure using your selected ISO/ASA film. Light reading by the camera's photocell is taken through the lens (TTL), rather than by the flash's light censor.

Cameras that have shoes for dedicated electronic flash have additional contacts on the hot shoe. The number of additional contacts depends on the number of functions that can be electronically performed by the flash/camera.

Do not use flashes dedicated for one camera on another brand; damage can result. Dedicated flash units on non-dedicated cameras must have all exposures set manually. If in doubt, check with your dealer.

Depth of field The amount of the picture in acceptable focus. In a scene that has depth it indicates how much of the total picture in addition to your focused subject you want in sharp focus. For example, with a f 2.8 aperture little of the total scene other than the subject is sharp. The background will be soft, hazy, and pastel-looking. With f 16, almost all of the total scene is sharply focused. Depth of field varies according to focal length of lens, subject/camera distance, and f stop used for particular lighting conditions.

DIN The European number designation indicating film sensitivity to light. The American designation is ASA. The International Standards Organization designation is ISO. ISO has now been adopted by the United States.

Electric eye Photosensitive cell used to determine the amount of light adequate for a particular film.

Electronic flash A flash unit that provides a great number of flashes, unlike flashbulbs where each picture requires a new bulb. Cost per flash is low. Electronic flash has a very short duration of light that floods the picture and freezes the action, usually at about 1/1000 of a second. There are units that operate on battery only, or battery/AC. Once fired, the flash unit usually needs a few seconds to recharge before being used again.

Emulsion The coating on the film that is light sensitive and on which the image is registered.

Existing light (See Available light.)

Exposed film Film that has been used to take pictures or that has been exposed to light in some other way. Opposite of unused or new film.

Exposure Aperture and shutter speed settings at which you shoot the film, such as f 16 at 1/60 second.

Extender arm Also called extender. For flash use. An arm mounted on the camera on which the flash unit is attached. Used to extend or increase flash distance from camera lens, and thus prevent red eyes in processed color pictures, white eyes in black and white pictures.

f stop Scale on a 35 mm camera lens that measures the size of the lens opening. "Stop down" or close down means to make the hole smaller. (See also Aperture.)

Focal length The length of the diagonal of the film image for a normal lens. It determines the lens's field of view. Also, the distance from the lens to the film.

Hot shoe The place on the camera where a flash is mounted and contact is made between camera and flash. Some shoes are *not* hot. They are simply for mounting, not connecting the flash. A separate cord must then be used to connect the flash to the camera.

ISO The current standard for indicating a film's sensitivity to light. Many boxes of film are labeled ISO/ASA. The higher the ISO number, the more sensitive the film is to light, and the less light is needed. When lighting conditions are poor, use film with a high ISO number. You *must* set the ISO on all cameras, except most instant loads. Sample ISO numbers: 25, 50, 64, 100, 200, 400, 1000, and so on. (See also Din.)

Kelvin Scale used to measure the color temperature of light. It has nothing to do with heat as expressed in Fahrenheit or Centigrade. When appropriate film is used for particular lighting conditions, no unusual color cast is visible on the processed picture. (See Color balance.)

Leader Film at the beginning of the roll you buy. For 35 mm, it is used to wind film onto the camera spool. It has holes on only one side of the film, unlike the rest of the roll of film which has holes on both sides.

Lens speed The maximum size opening (or f stop) to which a particular lens will open. The wider it will open, the "faster" the lens, and the more light reaching the film.

Fast Lens	Slow Lens
Lens opens more	Lens opens less
1.4, 2, 2.8,	4, 5.6 8, 11, 16

Many instant load cameras have a slow, fixed lens speed so flash is necessary in dim light. Not to be confused with shutter speed.

Light meter (or exposure meter) Built-in device in the camera or a separate hand-held unit used to measure the brightness of light for a particular subject. Designates the appropriate exposure for a particular ISO (film speed). It incorporates a light-sensitive cell to measure light, then computes appropriate shutter speed and aperture combinations. Found built in on most 35 mm adjustable cameras.

Manual override Term applies to an automatic/adjustable camera when the automatic feature is cancelled, and you can set both various shutter speeds and apertures yourself. This feature is useful for certain lighting conditions or to achieve some special photographic effects.

Overexposure Too much light reaching the film in the camera when shooting the picture. Before shooting, check your light meter where available. Also allow for special lighting conditions. See Figure 8–3 and InDepth IX (page 72). After the film has been processed, you can tell overexposure if:

1. The negative looks too "dense."
2. The print looks too light.
3. The slide looks light or washed out.

Parallax Difference in the viewing angle between the viewfinder and the lens. (See Rangefinder.)

Pushing film On 35 mm cameras, setting the camera ISO speed higher than the value indicated on the film box. This artificially increases the sensitivity of the film to light. You must note this increase when sending the film to be developed. The film will be processed by hand. You often pay an extra charge for this service. Not all films should be pushed, and the procedure is not generally recommended.

Rangefinder Found on instant load and on some 35 mm cameras. A viewing system where the viewfinder is separate from the lens. In looking through the viewfinder, you do not see exactly what the lens sees. However, you will include your entire desired subject if you compose the picture within the luminous frame or window in the viewfinder. Parts not viewed *within the luminous frame* will be chopped off in the processed picture. This problem becomes more evident in closeups. (See also Single lens reflex.)

Shutter speed How fast the camera shutter or curtain opens and closes to let light hit the film; therefore, how long film in the camera is exposed to light. You can adjust the shutter speed except for instant load and some automatic 35 mm cameras. Shutter speed of 1/30 second or slower may result in blurred picture. Place camera on tripod. A fast shutter speed can freeze motion so moving objects look clear. 1/1000 second will freeze motion of a moving train. Used in conjunction with aperture settings (f stop). Sample numbers on camera shutter speed dial: B, 1, 2, 4, 8, 15, 30, 60, 125, 250, 500, 1000.

Single lens reflex (SLR) Found on many 35 mm cameras. A type of camera with a viewing system that allows the subject to be seen directly through the picture-taking lens rather than through a separate viewfinder. You see exactly what the lens sees. With this viewing system, the lens that comes with the camera can often be replaced with special lenses, such as wide angle or telephoto for special shooting situations. (See also Rangefinder.)

Sprocket holes Holes on the edges of film that fit into camera sprockets to help insure proper advance and rewind of the film when it is in the camera.

Sprockets Teeth that engage the film sprocket holes to advance or rewind film in the camera. Sprockets are located on the inside of the camera where the film is loaded/unloaded. (See also Sprocket holes.)

Stop down See f stop.

Strobe Commonly but mistakenly used as a synonym for a simple electronic flash. A trade name.

TTL Through-the-lens. Often applied to metering.

Tungsten lighting Household lamps and special photo floods. Use tungsten film.

Underexposure Not enough light reaching the film in the camera when shooting a picture. Before shooting, cameras that have light meters or warning lights, either in the viewfinder or on the outside of the camera, will advise when there is insufficient light. See also Figure 8–3 and InDepth X (page 72). After film has been processed, you can recognize underexposure if:

1. The negative looks too "thin" (very light).
2. The print looks too dark.
3. The slide looks too dark or muddy.

Viewfinder A device in the camera that shows you the scene covered by the camera, and helps you to compose your picture. In many cameras the viewfinder can provide much information about camera functions, including f stop and shutter speed readings, and flash-ready light.

9 Visualmakers

CONTENTS

Operating Tips 78
   Camera Tips 78
   Electronic Flash Tips 79
Problems 79
InDepth
   Procedures for Using the
      Visualmaker 81
   Loading and Unloading Film
      in Camera 81
Attaching and Preparing Electronic
   Flash Unit, After Camera Has
   Been Mounted on Copy Stand .. 81
Batteries 82
Picture Size That Can Be
   Photographed with Copy Stand . 82
Hand-Held Shots with
   Copy Stand 83
What's Available in Film 83

The visualmaker is a copying kit used to make photographic slides or prints from any flat copy or small three-dimensional object. It is designed for Kodak Instamatic cameras. Some of the materials that can be reproduced well are prints, drawings, rock and mineral formations, graphs, passages from books, flowers, and small animals. Do not copy reflective or nonreflective glass-covered surfaces (they will glare) or those made of acetate. The kit comes with:

1. Portable copy stands in two sizes (3 inches by 3 inches, and 8 inches by 8 inches).
2. 126 format camera.
3. Pistol grip.
4. Electronic flash unit and assembly (old kits used flash cubes).

The copy stand can be operated on a tabletop or can be hand held to photograph closeup scenes in nature. When not using the copy stand, the camera can be operated like any ordinary Instamatic camera.

OPERATING TIPS

I. **Camera tips**

   A. Keep fingers and hands out of the picture area.

   B. Lighting for the photograph comes from a flashcube or electronic flash. Use either with the copy stand even when photographing outdoors.

   C. You do not need to look through the camera viewfinder when the camera is properly mounted on a copy stand. The subject being photographed is framed by the stand.

D. The film advance lever must advance until it locks; otherwise the shutter release will not depress.

E. If the film has not been completed and yellow paper still shows in the camera film counter, you may damage part of your film when you remove it from the camera.

F. Removing and replacing a partially exposed cartridge may result in the loss of at least one exposure.

II. Electronic flash tips

A. New batteries should last for about seventy-five flashes.

B. There is a faint humming when the flash unit is on. This is normal.

C. If you are not going to take pictures for at least fifteen minutes, move the RELEASE/OFF/ON switch to OFF.

D. Store the flash unit with the RELEASE/OFF/ON switch in the OFF position.

E. Remove batteries if unit is not to be used for a long time.

PROBLEMS

PROBLEM	CAUSE	SOLUTION
Camera Problems	*While Shooting*	
1. Film will not advance.	End of film.	Remove film.
	Film already advanced.	Shoot film, press shutter, and release.
2. Cannot press shutter release.	Film not fully advanced.	Advance film lever all the way.
Flash Problems—Magicube		
3. Cube will not fire when shutter triggered.	Dead flash.	Rotate cube a quarter turn to unused side or replace cube.
	Defective camera.	Take to dealer.
4. Red warning curtain appears.	Used cube.	Turn cube to another side or replace cube.
Flash Problems—Electronic Flash		
5. Electronic flash does not fire.	Improper connections.	Check camera and flash connections.
	Flash not ON.	Turn RELEASE/OFF/ON switch to ON.
	Weak batteries.	Replace.
	Batteries reversed in flash.	Check battery diagram inside battery compartment.

80 *Visualmakers*

PROBLEM	CAUSE	SOLUTION
Camera Problems	*While Shooting*	
	Battery compartment on flash not closed.	Close compartment.
	Ready light has not appeared.	Wait for ready light to come on.
	Broken flash cord. Defective flash unit. Defective camera.	Take to dealer.
6. Electronic flash will not fire, after having just been fired.	Unit needs to recharge.	Wait for ready light to come on.
	When Pictures Have Been Processed	
7. No picture.	No film in camera.	Put film in camera.
	Film did not wind through camera, or finished cartridge used.	Check loading.
8. Part of desired subject cut from picture.	Subject larger than copy stand frame.	Place copy stand frame on desired portion of total subject or use larger copy stand. (See InDepth V, VI.)
9. Pictures have incorrect colors.	Processing lab error.	Return negatives for reprinting.
10. Blurred pictures.	Out of focus.	Position camera lens carefully over copy stand lens.
	Picture-taking lens of camera dirty.	Wipe with camera lens tissue or spray tissue with liquid lens cleaner and wipe lens.
11. Glare	Acetate, glass-mounted subjects used.	Use subjects that do not cause glare, or use dulling spray on subjects. Spray is available from local art supply store.

IN-DEPTH

I. Procedures for using the visualmaker

A. Prepare the camera—load and advance the film, set the focus to beyond 6 feet (lens on copy stand will provide proper focus for closeup materials).

B. Decide which size copy stand to use. If using a large copy stand, open it until it locks into place.

C. Place the camera on the stand so that the camera lens fits into the copy stand closeup lens. Secure camera with the screw.

D. Insert the magicube or prepare the electronic flash unit. (See InDepth III.)

E. Arrange the material within frame of the copy stand. (See InDepth V.)

F. Press the camera shutter release.

G. Advance the film lever until it locks.

H. For electronic flash, wait for the ready light to come on. For magicube, replace it after you have taken four shots.

I. Take the next picture.

J. For electronic flash: when you have completed your picture taking, turn the RELEASE/OFF/ON switch to OFF. *Note:* The ready light is on; it will remain on for a few minutes.

K. To remove the flash unit from the copy stand, press the RELEASE on the flash unit and gently remove it from the stand.

II. Loading and unloading film in camera

Load the film as you would in any instant load camera. To unload after exposing the last picture, advance the film until all of the yellow paper disappears. Then remove the cartridge.

III. Attaching and preparing electronic flash unit, after camera has been mounted on copy stand

See Figure 9–1 (page 82).

A. Push the RELEASE/OFF/ON switch on the side of the flash toward RELEASE so that the mounting bracket extends.

B. Snap the large end of the cable assembly to the bottom of the flash unit.

C. Slide the cable assembly along the flash track until the assembly touches the bracket.

D. Insert the long end of the flash unit bracket into the top of the stand at the notched area of the stand. You may have to press down firmly.

E. Hold the flash unit with one hand, and with the other squeeze the bottom of the bracket (the end nearer to you) toward the flash unit. The bracket will retract so that the unit is clamped firmly onto the stand.

F. Insert the small end of the cable assembly at the top of the camera where the flashcube would ordinarily be placed. This completes the flash-camera connection.

Figure 9-1 *How to attach a flash unit to a visualmaker. (Courtesy Eastman Kodak Company)*

G. Turn the RELEASE/OFF/ON switch to ON. Wait for the flash unit ready light to come on. Cock the shutter, then take your picture.

IV. **Batteries**

A. Types of batteries:
1. Camera—batteries are needed in the camera for the electric eye for use off the copy stand. This is a round, 3-volt, PX 30 battery. The battery compartment on the camera is often separate from the film compartment. Turn the screw on the bottom of the camera to open.
2. Electronic flash—batteries needed to power the electronic flash unit, where applicable. The battery compartment is in the front part of the flash unit below the reflector—push up to open door; does not open all the way. Use only alkaline batteries. Replace batteries in the flash unit when it takes 30 seconds or more for the ready light to appear after the unit has been initially turned ON, or after the unit has been fired.

B. Cleaning battery terminals:
1. Use a rough cloth to clean battery ends. Use a pencil-type typewriter eraser to clean the contacts in the camera and in the flash unit battery terminals.

V. **Picture size that can be photographed with copy stand**

A. For the 3-inch-by-3-inch stand, 2¾ inches square is the maximum area you know will appear in the picture.
B. For the 8-inch-by-8-inch stand, 7½ inches is the maximum area that you know will appear in the picture.

Do not trim the extra material. Make sure that extra material in the subject being photographed extends beyond the sides of the stand, if possible, so that your photograph does not show an undesired black border. If the subject is much smaller than the stand,

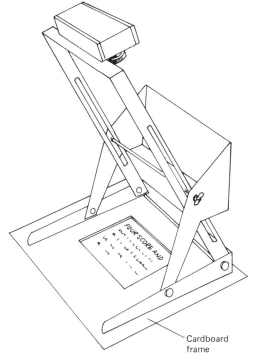

Figure 9-2 *Use cardboard frame for small items.*

use a matte cardboard frame at least the size of the stand. Center the picture on the cardboard. (See Figure 9–2).

VI. Hand-held shots with copy stand

The camera copy stand combination can be used off a table and hand held, pointed at an item, such as flowers in a vase. For closeups, the camera is not used without the copy stand, because the copy stand adds a closeup lens to allow shooting to within a few inches of the subject. Usually, you should hold the copy stand so that the open end of the frame is pointing up. Keep your fingers out of the picture area and out of the light path. Attach the pistol grip when using the large stand.

For each size copy stand, note the area of the picture that will be in clear focus when the stand is hand-held, and position your subject accordingly within the stand:

3-inch-by-3-inch stand: ½ inch above bottom of frame, ⅜ inch below.

8-inch-by-8-inch stand: 2 inches above bottom of frame, 1½ inches below.

VII. What's available in film

Instant load 126 film can be used both indoors and outdoors. It comes in twelve or twenty-four exposures for prints, and twenty exposures for slides.

Table 9–1

FILM	B&W/COLOR	PRINT/SLIDE
Verichrome Pan	B & W	Print
Kodacolor II	Color	Print
Kodachrome 64*	Color	Slide
Ektachrome 64*	Color	Slide

*Kodachrome 64 and Ektachrome 64 can be used interchangeably under similar lighting conditions. Note the differences below.

1. Kodachrome pictures usually have warmer tones, emphasizing reds. Ektachrome usually produces cooler pictures, emphasizing blues. This is a taste preference.
2. Kodachrome takes more time to process because it is sent away to Kodak, while Ektachrome can be processed locally.

10 Copy Stands

CONTENTS

Operating Tips 84
InDepth
    Homemade Copy Stands 84
    Lighting the Copy Stand 84

A copy stand is a device used to steady a camera when taking pictures of small three-dimensional objects or when photographing flat materials, such as the pages of a book. You may shoot prints or slides of the material being photographed, depending on the film used. You may shoot animations if you are using a movie camera (see Movie Cameras, Glossary).

The stand comes with or without lights and is fairly expensive. It consists of a flat piece of wood or masonite, say 16 inches by 20 inches, in back of which is a vertical metal post to mount the camera. The height of the camera is adjustable along the post to be as close to the subject as the camera lens permits, or as far away as the height of the post.

OPERATING TIPS

Always cover the flat surface on which you place the material to be photographed with gray or black cardboard or cloth.

INDEPTH

I. Homemade copy stands

If you do not have a copy stand, rig up your own in one of the following ways. (See Figure 10–1).

  A. Mount the camera to a bracket on wall. Place the subject on a flat surface below.

  B. Mount the camera on a tripod. Tack the copy against the wall.

  C. Use a tripod with the camera screwed in face down. Put the copy on the floor or other flat surface beneath the tripod.

II. Lighting the copy stand

  A. Daylight
    If the photography takes place outdoors in fairly bright light, no additional lights are needed.

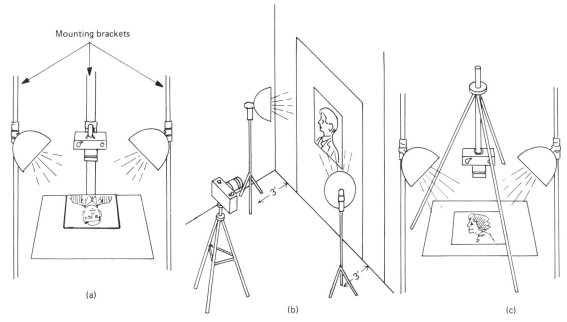

Figure 10-1 *Make your own copy stand.*

B. Artificial light. Use photo flood bulbs.

Photo flood bulbs are available at a camera dealer. Purchase or borrow reflectors in which to insert the bulbs. Make sure the bulbs are appropriate for the color film being used. Ask your dealer.

C. Where to place photo floods: two photo floods are usually used, one on either side of the copy stand, tripod, or other camera mount.

The floods must be about 3 feet away from and at a 45-degree angle to the subject being photographed.

11 Slide Projectors

CONTENTS

Operating Tips 88
Problems 88
   Power Problems 88
   Slide and Projector Problems...... 89
   Slide/Sound Synchronization
      Problems 92
   Sound Problems................. 93
InDepth
   Inserting Slides in Tray........... 93
   Procedure for Using Slide
      Projectors.................... 94
   Focusing....................... 94
   Projection of Slide Out of Tray
      Sequence (random projection)
      94

Projection of Single Slides Without
   a Tray 94
Trays........................... 94
Optional Equipment Useable
   With Many Conventional
   Front Screen Slide Projectors
   (not built-in screens) 96
Special Procedures for Advancing
   Slides...................... 96
Continuous Slide Projection 98
Cleaning........................ 99

Glossary 99

The slide projector is a very convenient piece of equipment, not only for showing commercially available programs but also for projecting homemade shows. Unlike filmstrip presentations, a slide program can be kept up to date by the replacement of individual slides.

    Slides can be made with almost any camera using film designated for that purpose. For example, usually color film followed by the term *chrome*, such as Kodachrome, will make color slides. If you have prints that you want to project, slides can be made from the print negatives.

    The most common slide formats are 35mm and 110 format (rectangular slides used horizontally or vertically), 126 format square slides, and super slides (large square slides). All of these will fit the standard slide projector in 2-inch-by-2-inch mounts or frames. The mounts are made of various materials, such as cardboard, glass, metal, or plastic, each having its own advantages and disadvantages.

COMMON TYPES AND FEATURES OF SLIDE PROJECTORS

Slide projectors for front screen operation are useable for small or large groups. Some have the following features:

1. Automatic focus.
2. Remote control capability for forward/reverse/focus.
3. Timer that allows for continuous projection at regular intervals.
4. Built-in sound capability.

Combined slide/sound projector units with built-in screens are useable primarily for individuals or small groups. (See Figure 11–1).

Sound portion:

1. Available as recorder/players or as playback machines only.
2. Controls like those of a cassette/recorder.
3. User allowed on some recorder/players to insert signal for automatic advance of slides at specified intervals.
4. Sight and sound to run continuously with the use of a special loop cassette.
5. Jack to add a headphone, or jack box for several headphones.

Slide portion:

1. Controls like silent slide projector.

Figure 11-1 *All-in-one slide-sound projector unit with built-in screen. (Courtesy of Telex Communications)*

2. Automatic focus.

3. Remote control capability for forward/reverse.

4. Some convertible from built-in to front screen projection with the flip of a switch. With front screen projection, the unit may be used for large groups.

OPERATING TIPS

1. Remove the cord and all accessories from the storage compartment before using the projector. This allows proper air circulation.

2. Keep your fingers off the slide; handle it by the frame only. Store away from heat.

3. When using REVERSE on the projector or on the remote control unit, depress button firmly; otherwise the slide tray may rotate forward.

4. The projector has a HIGH/LOW lamp switch. Use LOW most often. Use HIGH when the slides appear dim in a partially darkened room. Using the same lamp, HIGH gives 50 hours lamp life, LOW gives 200 hours.

5. When not using remote control, use FORWARD or REVERSE on the machine itself.

6. When using a slide/tape program that must be advanced manually, practice the presentation ahead of time.

7. When using a slide/tape program with automatic advance, ensure that slides and tape start at the proper place. View the introductory slide or hear the tape for instruction.

8. Whenever possible, use an 80-slot tray rather than a 140-slot tray. Slides in a 140-slot tray can jam easily.

9. Do not use cardboard mounts for slides, whenever possible. These mounts damage easily. Use glass or plastic.

10. For 110 slides:
a. Convert small mounts used for 110 slides a few years ago to 2-inch-by-2-inch mounts by using special inexpensive mount adaptors on each slide.
b. Move the projector away from the screen toward the back of the room to increase 110 picture size. Of course, you will lose some picture quality as the image size increases.

PROBLEMS

PROBLEM	CAUSE	SOLUTION
Power Problems		
1. No power.	Cord not plugged into machine or outlet.	Check connections.
	Cord defective, frayed, or improperly connected.	Check connections.
	Defective machine.	Take in for repair.
2. Black screen.	No power.	Check power cord connection.
	Projector not turned on.	Turn to FAN, then to LAMP.

Slide Projectors 89

| | Lamp burned out. | Replace lamp after fan has cooled machine, and you have unplugged cord from AC outlet. |

Slide and Projector Problems

3. Tray does not fit into projector.	Top and bottom of tray improperly aligned.	Rotate metal bottom plate of slide tray until it locks slot on plate at "0" notch on tray. Set "0" of tray at gate index of projector.
4. Black screen.	Projector works. On projectors with black shutter, shutter remains closed when slide is missing.	Fill tray space with slide, if desired.
5. White screen, no picture, or only part of picture showing.	Slide in tray does not drop into projector at all, or drops in part way.	Press FORWARD or REVERSE. Otherwise, do as follows:
	• Slide mount frayed at corners.	Clip slide mount off slightly with nail clipper.
	• Slide mount warped.	Remount slide in new mount.
	• Tray not properly positioned.	Reposition tray. (See Problem 3 above.)
	No slide for that slot in tray. Projector does not have black shutter.	Insert slide.
6. Projector gets very hot; slides burning.	Heat filter lens missing or cracked.	STOP projector. Take in for repair or replace heat filter.
7. Fan runs; lamp does not light.	Projector switch on FAN.	Flip switch to LAMP.
8. Poor illumination.	Room not sufficiently darkened (front screen use).	Darken room or bring projector closer to screen.
	Lamp not fully seated in socket.	Remove lamp; note proper insertion.
	Lamp on LOW.	Turn to HIGH.
	Old, dark-looking lamp.	Replace lamp.
	Thin extension cord used.	Use heavy-duty cord.

Slide Projectors

PROBLEM	CAUSE	SOLUTION
Slide and Projector Problems		
	Dirty lens.	Remove lens. Clean both ends.
	Condenser lens broken.	Take in for repair.
9. Slide blurred.	Projector out of focus.	Focus from projector itself or remote control.
	Remote control focus defective.	Use FOCUS knob on projector.
	Image out of focus on slide.	Cannot improve picture. Discard slide and shoot again.
	Slide has fingerprints.	Clean slide carefully with lens tissue, not facial tissue.
	Some slide mounts are different thickness from other mounts.	Refocus particular slide. (See InDepth III, VI A.)
	Dirty lens.	Remove and clean lens.
	Heat filter lens missing or broken.	Replace heat filter lens.
10. Image upside down or backwards.	Slide inserted incorrectly.	Remove slide. Hold slide so it looks upside down and backward as you might face a front screen. Insert slide.
11. Lens barrel falls out of machine.	Lens barrel rotated too far to focus. Projector too close to screen, out of minimum focusing range.	Reinsert lens barrel until it clicks. Move projector away from screen. Use FOCUS knob.
12. Slide stuck in projector, does not pop back into tray.	Slide mount too thick or warped.	1. To pop slide back into tray: • press SELECT button, or • REVERSE. 2. To remove it from machine, use emergency tray removal procedures, then remove slide. (See InDepth VI, C 2.) 3. Remount slide.

Slide Projectors 91

13. Remote control does not work.	Remote unit faulty.	Check that remote is plugged in with colored dot up. Check remote control on another projector. Use FORWARD and REVERSE on machine.
	Defective solenoid on projector.	Try FORWARD and REVERSE on machine. If faulty, take in for repair.
14. Tray will not advance to next slide.	Tray not properly positioned.	Reposition tray (see Problem 3 above).
	Slide stuck in projector.	See Problem 12 above.
	Nubs on bottom of tray broken.	Push SELECT button and manually move tray past damaged area, or Use tray only up to damaged spot, or Load slides into new tray.
	Remote unit faulty.	Check that remote is plugged in with colored dot up. Check remote control on another projector. Use FORWARD and REVERSE on machine.
	Defective solenoid on projector.	Try FORWARD and REVERSE on machine. If faulty, take in for repair.
15. Remote control only works backward.	Remote control plug inserted into machine upside down.	Plug in with colored dot up.
	Remote FORWARD defective.	Check FORWARD on machine rather than remote control.
16. Tray will not reverse to previous slide.	REVERSE on remote control or projector not firmly depressed.	Firmly depress REVERSE.
	Slide stuck in projector.	See Problem 12 above.
	Tray not properly positioned.	See Problem 3 above.

92 *Slide Projectors*

PROBLEM	CAUSE	SOLUTION
Slide and Projector Problems		
	Remote REVERSE defective.	Use REVERSE on machine rather than on remote.
17. Slides fall out of tray when tray turned upside down.	Locking ring not in place atop tray.	Put locking ring in.

Slide/Sound Synchronization Problems (for built-in combination unit or separate cassette unit designed to synchronize tape with projector.)

PROBLEM	CAUSE	SOLUTION
18. Slides not advancing automatically.	Tape not programmed for automatic advance.	Advance manually.
	Automatic advance signal on that tape will not work on your machine.	Advance manually.
	When duplicated, automatic advance tape duplicated on one side only.	Duplicate both sides of original when tape programmed for automatic advance.
	Defective machine.	Check tape on another machine. Take in for repair.
19. Hear beeps only; tray does not advance.	Playing wrong side of tape.	Turn cassette to side 1. Play side 1 only.
20. Sound and picture not synchronized. For additional synchronization problems, see also chapter on Dissolve Unit, Problems, 21, 22.	Tape and slides not started simultaneously at same place.	Rewind tape to beginning. Move slide tray to "0" or "1" depending on whether sound starts at "0" or "1" slide. Press PLAY on sound portion. Slides will advance automatically or manually, depending on tape signals and machine capability. (See InDepth VIII.)
	Tape rewound or advanced without adjusting slide position.	Move slide tray to coordinate with tape.
	Slides forwarded or reversed without adjusting tape position.	Advance or rewind tape to match slides.

Sound Problems

See Chapter 18 for any of the problems dealing with sound listed below. Unlike slide/sound problems, these situations are not directly related to the functioning of your slides. You may have sound problems when using a self-contained slide/sound unit, or when using a separate tape recorder with your slide projector whether or not the recorder is designed for automatic tape advance.

- No power.
- No sound on playback (see also Problem 1 above).
- Weak sound on playback.
- Sound fading in and out on playback.
- Sound mushy, wow and flutter, distorted sound, noise on playback.
- Humming or scratchy noise on playback. On playback, notice that previously recorded material did not erase completely.
- RECORD button does not depress.
- No RECORD button on machine.
- Tape stuck.
- Tape squeal or squeak.
- Tape runs backward on AC.
- Tape unthreads and jumbles while machine is running.

IN-DEPTH

See Figure 11–2 for basic parts of a slide projector.

I. Inserting slides in tray

A. As you face the screen, put the slides in the tray upside down and reversed with lettering upside down and backward.

B. When holding the slides as above, mark them in the upper right-hand corner with a dot or an X so you will always insert them correctly.

Or dot the top right edge of each slide with a colored marker, after the slides have been inserted correctly.

Figure 11-2 *Front, back controls for Kodak carousel-type projector. (Courtesy Eastman Kodak Company)*

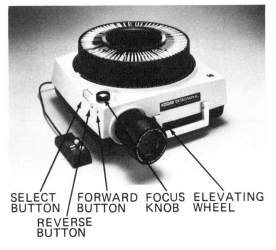
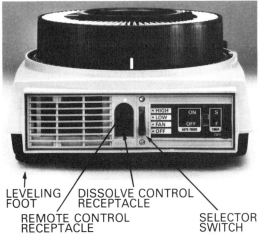

Or, if you have more than one tray, use one colored marking pen per tray—red for tray 1, blue for tray 2, and so on.

II. Procedure for using slide projectors

A. Remove the power cord from the storage compartment and unwind. Plug in the cord with the machine OFF.
B. Connect remote control, if available, with colored dot on plug facing up.
C. Install the tray with the "0" at the projector gate index. If it does not seat properly, check the metal bottom plate. (See Problem 3.)
D. If the machine has a manual advance option, set the timer to M.
E. Turn on the projector, moving the lamp switch to LOW.
F. Press FORWARD on the remote control to slide 1. If not using remote control, press FORWARD on the projector.
G. Focus the image.
H. If using zoom lens, turn the front of the barrel to make the picture larger or smaller on the screen.
I. Begin your presentation.

III. Focusing

Focus with the remote control or the FOCUS knob on the projector.

For automatic focus models, only the first image requires focusing. However, if using glass- and cardboard-mounted slides in the same presentation, you may need to readjust the focus for the different slide mounts.

For manual focus models, you may need to focus each slide as it is projected.

IV. Projection of a slide out of tray sequence (random projection)

A. Press and hold down the SELECT button.
B. Rotate the slide tray. The desired slide number should be at the projector gate index.
C. Release the SELECT button. The slide will be projected.

V. Projection of single slides without a tray

A. Insert the slide into the projector slot.
B. Use the SELECT button to remove the slide from the projector.

VI. Trays

A. Kinds of trays (Kodak circular slide trays, using 2-inch-by-2-inch frames are discussed below).
   1. 80-slide capacity comes in two types: eighth-inch-thick slide slots for all types of frames or mounts, or tenth-inch-thick slide slots for cardboard or thin glass mounts.
   2. 140-slide capacity comes in sixteenth-inch-thick slide slots for cardboard or thin plastic mounts.
      Whenever possible, use the 80-slide tray rather than the 140-slide tray. Slides in a 140-capacity tray tend to jam easily.
B. Parts of a slide tray and their functions.
   1. A slotted plastic tray area for placing slides is numbered 0-80

or 0-140. The "0" has a notched edge for placing the tray at the projector gate index. For normal use, slide spaces are numbered from "1" on. For continuous operation, see In-Depth IX.

2. The plastic locking ring on top of the tray can be removed counterclockwise to insert slides. Replace and lock it to keep the slides from falling out.

3. The metal plate under the plastic tray is for tray advance or reverse, random slide projection, and emergency tray removal. When turning the tray upside down to check the metal plate, keep the locking ring in place so that the slides will not fall out.

   The slot of the metal plate should be next to the "0" of the tray. If not, rotate the metal plate until it clicks.

C. Tray removal.

   1. For normal tray removal

      a. If the "0" is opposite the gate index of the projector, lift off the tray.

      b. If the "0" is not opposite the gate index, return it to "0" as follows:

         • Turn the projector ON, in most cases.

         • Press and hold down the SELECT button.

         • Rotate the slide tray in either direction, and return the "0" to the gate index.

         • Release the SELECT button.

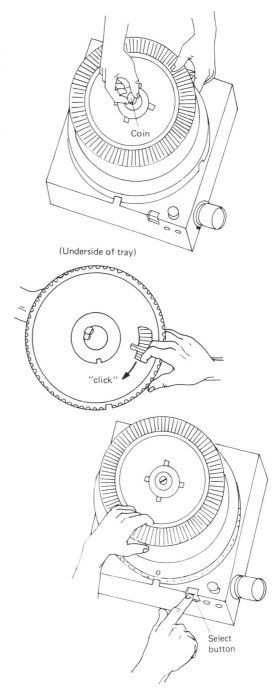

Figure 11-3 *Emergency tray removal and replacement.*

- Lift off the tray.
2. For emergency tray removal—without returning "0" to the gate index, such as when a slide is stuck in the projector—see Figure 11–3.
    a. Locate the large projector screw in the center of the tray.
    b. Turn the screw with a coin in either direction as far as it will go, and hold it turned.
    c. Tilt and lift off the tray. Release the screw.
    d. Correct the malfunction in the projector.
    e. Make sure the locking ring is holding the slides in place in the tray. Turn the tray upside down and rotate the metal plate until it clicks. Fit the tray back into the projector.
    f. Keeping the SELECT button depressed, rotate the tray to the last slide projected. Release the SELECT button.
    g. Resume presentation.

VII. **Optional equipment useable with many conventional front screen slide projectors (not built-in screens)**

   A. Dissolve unit. This device provides a smooth visual transition and/or special effects during slide changes. It usually requires at least two projectors and two slide trays, but only one screen. See chapter on Dissolve Unit.
   B. Lenses. There is a wide variety of lenses available. If your projection needs vary, select a zoom lens, where the size of the projected image can be adjusted without having to change the projector-to-screen distance.
   C. Special cassette recorder. When connected to the projector, this sound unit can automatically advance the slides at predetermined intervals.
   D. Filmstrip adaptor.
   E. Bulk load slide unit or stack loader, for quick review of up to forty slides without a tray.
   F. Remote control extension cord, 25 feet.

VIII. **Special procedures for advancing slides**

   A. Automatic advance of slides by projector timer, where available (no sound).

   Set the projector timer at specified intervals, where possible (as 5, 8, or 15 seconds) or flip the manual/automatic switch to automatic, where available. Automatic operation can be interrupted at any time by FORWARD or REVERSE on the remote control. When the timer is on an automatic setting, it can keep the machine rotating indefinitely until you stop it. This allows for a continuous slide showing.

   B. Automatic advance of slides by programmed tape.
      1. General information.
         a. Prerecorded tapes programmed to automatically advance slides at certain intervals have inaudible sig-

nals recorded at 1000 hz frequency. Commercial slide/cassette programs are standardized at this frequency.

b. A slide projector with built-in sound will activate 1000 hz signals. If you are using a separate silent slide projector, use special cassette equipment which synchronizes at 1000 hz frequency. Connect the two units with an appropriate patchcord recommended by your audio/visual dealer. No special sound equipment is necessary to play the taped narration when you do not want to activate the automatic signals.

c. Whether using automatic signals or just listening to the contents, always use side 1 only.

2. Procedure.

a. Set the projector timer to MANUAL.
b. Turn the machine ON.
c. Put the projector lamp switch on LOW, wherever possible.
d. Insert the cassette tape on side 1. Rewind to the beginning.
e. Insert the slide tray at "0."
f. Press PLAY on the cassette unit.
g. Listen to the tape instructions to determine whether the tape signals advance the slide tray beginning at the "0" or "1" position.
h. Reset the tray at "0" or "1"; rewind the tape to the beginning.
i. Begin the presentation by pressing PLAY on the cassette unit.
j. Stop the presentation when necessary, using STOP or PAUSE on the cassette unit.
k. If a slide gets out of sequence, stop the projector and the cassette unit. Use FORWARD or REVERSE on the projector to synchronize visuals with sound.

C. Manual slide advance coordinated with accompanying taped narration.

Set the projector timer to M (manual). To advance slides manually, use FORWARD/REVERSE controls on the projector or on the remote control unit, as appropriate when a cassette tape is playing. Where the tape is not designed to advance slides automatically, pictures and sound can be coordinated by:

1. Recording audible signals on the tape (such as using a clicker, clinking a glass or ringing a bell) during recording so you will know when to manually advance slides, or

2. Following a script during projection.

Some commercially available slide/tape programs have audible signals. These signals are your cue telling you to advance the slide projector.

D. Advance of slides using tape originally programmed for filmstrip projector.

1. Automatic advance.

   Tapes made to signal automatic advance of slide projectors and tapes made to signal automatic advance of filmstrip projectors have different frequencies (1000 hz for slides, 50 hz for filmstrips). If you convert a filmstrip into slides, the tape made for the filmstrip will not automatically advance the slide projector, unless you have a special cassette player equipped for both 1000/50 Hz use.

2. Manual advance.

   Tapes made for filmstrips often have the narration repeated on sides 1 and 2. Use the audible cue on one side of the tape to help you advance the slide projector manually. The other side has the inaudible automatic advance signal.

IX. **Continuous slide projection**

   A. For automatic slide advance, use a projector timer or a cassette programmed for automatic advance. (See InDepth VIII A, B.)

   B. Prepare a slide tray to make sure all slots appear filled. For continuous operation of the same program, no slide tray slot, including "0" space, should appear empty or white on the screen. Handle the "0" space as follows:

   1. Black shutter projectors. No special tray preparations are necessary. Recent slide projectors have a black shutter designed to remain in place when there is no slide in a given slot. The screen appears purposely black rather than empty.

   2. Projectors with no black shutter. Many slide projectors do not have a black shutter feature. During continuous projection the "0" space will appear empty. Since a slide cannot fit in the "0" space of the tray as usually inserted, fill the space as follows:

      a. Remove the tray when "0" is at the gate index. Insert a slide into the *projector* between the two metal guides. Replace the tray as usual.

      b. Begin projection. You will now have 81 instead of the normal 80 slides. The slide will enter the "0" space in the tray and will be projected. When the tray is returned to "0" at the end of the program, the slide will again drop into the projector.

   C. Tray removal.

      To remove the tray, rotate it so "0" is opposite the projector gate index. After lifting off the tray, if you used an extra slide be sure to remove the extra slide from the gate by pushing the SELECT button. Store the slide in the center of the tray.

   D. Sound. When necessary, use a continuous loop cassette designed for the number of minutes the show runs, usually up to 12 minutes.

X. Cleaning

A. Lenses for front-screen projectors. Clean the projection lens by removing it from the projector and using a soft lint-free cloth or lens tissue to clean both ends of the lens. Also clean the condenser lenses and the heat filter lens inside the projector.

B. For slides, use lens tissue and wipe carefully. Do not use facial or eyeglass tissue.

C. For built-in screens, use an art gum eraser or wash with mild soap on a soft moist cloth. Gently rinse with clean cloth and wipe dry.

GLOSSARY

Condenser lenses Lenses located inside the projector that concentrate light through the image to be projected.

Dissolve unit An accessory that allows gradual change of slide image and special effects on screen. (See Dissolve Units.) There are also all-in-one slide/sound/dissolve front screen projectors that do not require separate dissolve units.

Ektagraphic projector Professional model of the Kodak Carousel projector.

Gate index Notched area of projector which exposes the number on the revolving slide tray being shown at a given time. Normally the "0" on the tray must be opposite the gate index in order to insert or remove the tray.

Heat filter Special lens that prevents the heat of the projector lamp from burning the slide.

Remote control An accessory that lets the user advance, reverse, or focus the projector without being next to the projector. Some automatic focus projectors do not have a focus feature on remote control.

Select button or bar Feature found on the top or side of the projector. Necessary for:

- Random slide projection. (See InDepth IV.)
- Returning the tray to "0" for normal tray removal. (See InDepth VI C.)
- Retrieving single slides projected without a slide tray. (See InDepth V), and
- Removing a slide that is stuck in the projector and will not come back into the tray. (See Problems #12.)

12 Dissolve Unit

CONTENTS

Operating Tips 100
Problems 102
   Setting up Dissolve Unit 102
   Operating Dissolve Unit/Projectors 103
   Tape Recorder/Dissolve Unit/
      Projector Problems 106
   Sound Problems 107
InDepth
   Inserting Slides in Trays 107
Setting Up the Dissolve Unit for
     Manual, Live Operation 108
Setting Up the Dissolve Unit and
     Operating It Automatically or
     Continuously (no sound) 111
Setting Up the Dissolve Unit
     and Operating It
     Automatically with Sound ... 112
Glossary 112

A dissolve unit is used with slide projectors. It provides a smooth, professional-looking transition during slide changes by fading in and out slide images. Many programmable units can also create special effects. One image appears on the screen, although with most dissolve units two or more projectors are used. In addition to making a presentation more professional-looking, you can use many more slides without having to stop your presentation for a change of trays.

This chapter covers slide presentation procedures using a dissolve unit. It does not cover specific production procedures using a programmable dissolve unit.

OPERATING TIPS

1. Consider whether your special effects are to be performed live during the presentation or programmed ahead of time. Also decide whether you want to use sound. Consult your audio/visual dealer to evaluate the capability of your equipment.

2. Each slide projector must have the appropriate receptacle to accept a dissolve unit plug. Check with your audio/visual dealer, if necessary.

3. Most dissolve units require at least two slide projectors. Many units accept more.

The slide projectors can be used side by side or stacked one above the other on shelves called stackers.

4. A change in the order of slides with one projector requires a similar change with slides of the second projector, so the presentation remains coordinated.

5. For a smoother-looking, more consistent effect, you may want to select slides that have all been photographed either from the horizontal or from the vertical perspective.

6. On older projectors, the first slide and the last slide of the entire presentation should be black. (See Figure 12–1.) This provides an effective dissolve, fading in at the beginning and fading out at the end of the program.

7. Newer projectors have a black shutter that covers spaces with no slides (empty spaces). This eliminates the need for black slides.

8. Dissolve units can fade in and out images or cut from one image to another. The speed with which these effects occur can be controlled.

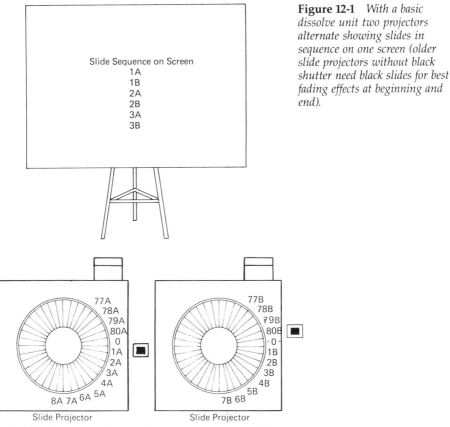

Figure 12-1 *With a basic dissolve unit two projectors alternate showing slides in sequence on one screen (older slide projectors without black shutter need black slides for best fading effects at beginning and end).*

With a basic dissolve unit, two projectors alternate showing slides in sequence on one screen.

PROBLEMS

PROBLEM	CAUSE	SOLUTION
Setting up Dissolve Unit		
1. Black screen, or black screen every other slide.	No power.	Check power cord connection.
	One or both projectors not turned on.	Turn projector to FAN.
	Lamp burned out on one or both machines.	Replace lamp after fan has cooled machine.
2. Tray does not fit in projector.	Top and bottom of tray improperly aligned.	Rotate metal bottom plate or slide it until it locks. Slot on plate locks at "0" notch on tray. Set "0" of tray at gate index of projector. Place in projector.
3. Poor illlumination. Check to see which machine seems to project least light.	Room not sufficiently darkened.	Darken room or bring projectors closer to screen.
	Lamp not fully seated in projector socket.	Remove lamp. Note proper insertion procedure.
	Old, dark-looking lamp.	Replace.
	Thin extension cord used.	Use heavy-duty cord.
	Dirty lens.	Remove lens. Clean both ends.
4. Images on the screen not the same size.	Lenses from each projector do not match.	Projector with smaller picture should be moved farther back than other projector, or match lenses on both projectors.
	Projectors using same lenses not equidistant from screen.	Even up projector distance from screen.
5. Images not centered over each other.	One projector facing too far right or left, or too high/low.	Adjust projector positions so that images from both overlap exactly.
6. Pictures fail to fade in and out; slides change abruptly.	Dissolve unit not properly connected.	Check dissolve-projector connections.

Dissolve Unit   103

	Dissolve unit set for CUT.	Change dissolve unit to FADE.
	Projector switch turned to HI or LOW lamp.	Set each projector's switch on FAN.
	Dissolve unit malfunctioning.	Continue presentation without unit as follows 1. Disconnect dissolve-projectors. 2. Plug AC cord of each projector into wall. 3. Set up an extra screen, one for each projector. Focus each projector on its screen. 4. Use separate control unit for each projector. 5. Alternately operate remote control of each projector.
	Projector(s) with permanently attached AC cord, plugged into AC outlet.	Plug AC cord(s) of projector(s) into dissolve unit.
7. No automatic advance of slides.	Where necessary, no synchronizing cable connected from dissolve unit to audio tape player.	Connect appropriate cable.
	Defective signals, tape.	Check that signal frequency is compatible with your equipment and tones are properly spaced. Check that tape is not damaged.
	On old units, dissolve unit on MANUAL.	Switch dissolve timer to desired interval setting.

Operating Dissolve Unit/ Projectors

8. White screen—no picture or only part of picture showing.	Slide in tray does not drop into projector at all, or drops in part way.	Press FORWARD or REVERSE. Otherwise, do as follows:
	• Slide frayed at corners.	• Clip off slightly with nail clipper.

104 *Dissolve Unit*

PROBLEM	CAUSE	SOLUTION
Operating Dissolve Unit/ Projectors		
	• Slide frame warped.	• Remount slide in new frame.
	• Tray not properly positioned.	• Reposition tray. (See Problem 2 above.)
	No slide for that slot in tray.	Insert slide in projector that lights up when screen is white. (See also Slide Projectors, InDepth IX B).
9. Tray will not advance to next slide.	Tray not properly positioned.	Reposition tray. (See Problem 2 above.)
	Slide stuck in projector.	See Problem 10.
	Faulty remote unit attached to dissolve unit.	Check that remote is plugged into dissolve unit with colored dot up.
		Check remote control on another projector.
		Abandon remote. Use FORWARD and REVERSE on each machine.
	Sync cable, signals or tape problems.	See Problem 7.
10. Slide stuck in projector, does not pop back into tray.	Slide frame too thick or warped.	To pop slide back into tray, press SELECT button or REVERSE.
		To remove it from machine, use emergency tray removal procedures, then remove slide. (See Slide Projectors, InDepth VI C 2.)
		Remount slide.
	Slides have been placed in 140-slide trays.	Trays holding 140-slides are not recommended for multi-projector, multi-image presentations. Use only 80-slide trays.

Dissolve Unit **105**

11. Cannot back up slides.	Check REVERSE capability of your dissolve unit or remote control.	In some cases use REVERSE on projectors, not remote.
	Slide stuck in projector.	See Problem 10.
	Tray not properly positioned.	See Problem 2.
12. Every second slide blurred.	One projector not focused properly.	Projector that lights up when out-of-focus slide is projected needs focusing. Use FOCUS knob on projector.
	Dirty lens on projector that lights up.	Remove lens and clean.
13. Slides blurred randomly.	Some slide mounts are a different thickness from other mounts.	Refocus particular slides, using FOCUS knob on projector. (See Slide Projectors, InDepth III, VI.)
	Where available, FOCUS defective on remote control attached to dissolve unit.	Use FOCUS knob on projector.
	Image out of focus on slide.	Cannot improve picture. Discard slide and shoot again.
	Where special switch available, projector AUTO FOCUS off.	Turn AUTO FOCUS on.
	Slide has fingerprints.	Clean slide carefully with lens tissue, not facial tissue.
14. FOCUS does not work on remote control attached to dissolve unit.	Some projectors have automatic focus; remote FOCUS will not have effect.	Use projector FOCUS knob for out-of-focus slide. Other slides will focus automatically. (See also Problems 12 and 13.)

106 *Dissolve Unit*

PROBLEM	CAUSE	SOLUTION
Operating Dissolve Unit/ Projectors		
15. Remote control attached to dissolve unit does not work.	Defective remote control, or no remote capability on dissolve unit.	Check remote control by: 1. Operating it on another projector. 2. Unplugging remote control from dissolve unit; operate dissolve unit on an automatic cycle.
16. Remote control attached to dissolve unit works backward.	Remote control plug inserted into dissolve unit upside down.	Plug in with colored dot up.
	Remote FORWARD defective.	Use FORWARD on machine rather than on remote control. Replace remote control.
	Defective dissolve unit.	Take dissolve unit in for repair.
17. Slides fall out of tray when tray turned upside down.	Locking ring not in place atop tray.	Put locking ring in.
18. Lens barrel falls out of machine.	Lens barrel turned too far to focus.	Reinsert lens barrel until it clicks. Use FOCUS knob. You may need to move projector away from screen for minimum focusing distance.
19. Projector gets very hot; slides burning.	Heat filter lens missing or cracked.	STOP projector. Take in for repair or obtain and replace proper filter.
20. Image upside down or backward.	Slide inserted incorrectly.	Remove slide. Hold slide so it looks upside down and backward as you face screen.
Tape Recorder/Dissolve Unit/ Projector Problems		
21. Tape emits synchronizing signal; projectors on, but not moving.	Dissolve unit ON/OFF is OFF.	Turn dissolve unit ON.

	Special cord from recorder not plugged into dissolve unit.	Plug recorder cord into dissolve unit.
	Plugs from dissolve unit to projectors not in projectors.	Check connections.
	No apparent reason.	Try to flip dissolve unit OFF/ON once. (See also Problem 22, cause 1.)
22. Projectors do not fade in and out properly according to tape signal.	Wild, erratic behavior because more than one AC outlet used.	Plug *all* equipment into the *same* AC outlet. Use multiple outlet strip, if necessary.
	Dissolve unit timer where available, and/or projector timers set on automatic.	Set all timers to MANUAL.
	On occasion, signals are programmed too close to each other.	Reprogram signals. Allow sufficient time lapse between each signal so that dissolve unit and projectors can complete their functions before the next signal.

Sound Problems

For tape recorder sound problems below, see chapter on Tape Recorders.

- No power.
- No sound.
- Weak sound.
- Wow and flutter, distorted sound.
- Fading in and out.
- Humming or scratchy noise.
- Mushy sound; no high frequencies.
- Unwanted radio station programming.
- Tape stuck; will not move.
- Tape squeal or squeak.
- Tape unthreads and spills out of housing.
- Tape not automatically synchronizing other equipment.

IN-DEPTH

I. Inserting slides in trays

With an ordinary slide presentation, you use one projector, therefore one slide tray. With the dissolve unit, you use two or more projectors, therefore two or more slide trays. Load slides in trays in order in which the slides will appear on the screen. Note the sample sequence for inserting slides. See Figure 12–1, page 101. While the slides are viewed in sequence, each of two or more projectors is taking turns showing the slides in its own tray. For easy reference, consider only two projectors. Label the left tray A and the right tray B. Label slides 1A, 1B, 2A, 2B, 3A, 3B, and so on.

First slide of show—left projector tray first slide—1A

Second slide of show—right projector tray first slide—1B

Third slide of show—left projector tray second slide—2A

Fourth slide of show—right projector tray second slide—2B

II. Setting up the dissolve unit for manual, live operation

Equipment needed: Two slide projectors, each using the same type of lens (except when a zoom lens is used), one remote control unit when compatible with your equipment and where desired, one dissolve unit, one screen.

A. Setting up

1. Remove all cords and remote control units from cord compartments of projectors.
2. Place carousel projectors together, with dissolve unit to one side or below projectors, if you are using cart with shelves.
3. Place trays in each projector with "0" at gate index.

Figure 12-2 *Three ways to operate slide programs with dissolve unit, after making basic connections.*

Method A	Method B	Method C
HAND OPERATION LIVE, NO SOUND	CONTINUOUS AUTOMATIC OPERATION AT REGULAR BUILT-IN INTERVALS. LIVE, NO SOUND	AUTOMATIC OPERATION, SPECIAL EFFECTS; PROGRAMMED AHEAD OF TIME, SOUND
Connect Dissolve unit to Remote control unit, if desired.	Connect Basic dissolve-projectors connections.	Connect Dissolve unit to synchronizing tape recorder.
With older dissolve units Timer on MANUAL Use remote control	With older dissolve units Timer on AUTOMATIC Where possible set dissolve switch on desired interval.	With older dissolve units Timer on MANUAL Activate tape recorder and dissolve unit.
OR	OR	OR
With newer dissolve units With remote control, select basic forward/reverse functions on remote unit. Without remote control, select desired special effects on dissolve unit during presentation.	With newer dissolve units Plug in optional external timer module, if available, and dial desired automatic intervals on timer.	With newer dissolve units If connected, disengage external timer. Activate tape recorder and dissolve unit.

4. Insert slides. (See InDepth I.)
5. Connect equipment. (See Figure 12–3.) All AC connections for dissolve system must be on the same phase and ground; otherwise, the system will be erratic. Use one wall outlet, with a multiple outlet strip, if necessary.

    Plug the dissolve unit power cord into the wall outlet. Plug dissolve-projector cords (7-pin plugs) into outlets in back of each projector. If using projectors with permanently attached power cords, plug cord of each projector as shown.
6. For manual remote operation at a live presentation, connect remote control to dissolve unit, where possible. Remote plug colored dot normally faces up.
7. Set projector timers to MANUAL.
8. Determine how to activate dissolve function manually on your unit. (See Figure 12–2, Method A.)
9. Flip ON/OFF switch of dissolve unit to ON, or otherwise activate unit.
10. Focus and adjust the screen image. Coordinating the images is not difficult. However, if you do not achieve the desired results, retrace your steps from the beginning.

    Both projectors—Turn switches to FAN.

    Left projector—Press FORWARD on the remote control of the projector or CYCLE or ADVANCE on the dissolve unit, until the left projector lights up. On older projectors, the screen lights up without a slide. On models with the black shutter, a slide will trigger black shutter to rise, and allow light to reach the screen. Use a registration slide in your tray to align and focus projector. Otherwise, press FORWARD on the projector until a slide appears on the screen horizontally rather than vertically. Center the picture on the screen.

    Focus the slide, using the remote control FOCUS or FOCUS knob on the projector. Remote control units for automatic focus projectors may not have FOCUS control.

    Adjust the projector height and tilt.

    Adjust the image size on the screen by changing projector-to-screen distance or rotating zoom lens where available.

    Flip lamp switch to LOW.

    Right projector—Use the same procedure as for the left projector: Press FORWARD on remote control of the projector or CYCLE or ADVANCE on the dissolve unit until right projector lights up. On older projectors, the screen lights up without a slide. On models with the black shutter, a slide will trigger black shutter to rise, and allow light to reach the screen. Use a registration slide in your

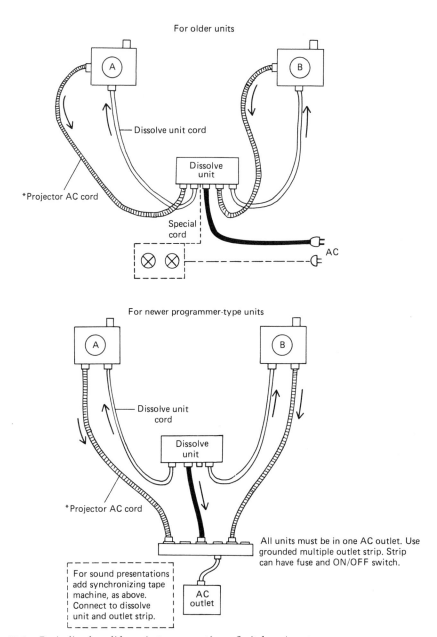

Figure 12-3 *Basic dissolve-slide projector connections. Switch ON/OFF to FAN on projectors. All AC cords must plug into only one wall outlet. (*Connection only necessary where slide projector has built-in AC cord.)*

tray to align and focus projector. Otherwise, press FORWARD on the projector until a slide appears on the screen horizontally rather than vertically. Center the picture on the screen.

Focus the slide, using the remote control FOCUS or the FOCUS knob on the projector. Remote control units for automatic focus projectors may not have focus control.

Match height, leveling, and image size on the screen with the left projector until the slides overlap perfectly to form one image on the screen.

Left projector—Return the switch to FAN.

Right projector—Check the focus of the image on the screen. The projector should remain on FAN.

Both projectors—Use the SELECT button on each projector to release the tray and return each tray to "0."

B. Showing your presentation, using manual operation.

As noted above, if possible with your equipment and if desirable, use remote control plugged into dissolve unit. Press FORWARD on remote control and focus when appropriate. Test before your performance. Where you cannot REVERSE remote or dissolve unit, use projector REVERSE, if necessary.

C. After the presentation:
1. Switch the dissolve unit timer to MANUAL, where available.
2. Return each projector tray to "0" using SELECT on each projector; some dissolve units are equipped to automatically return projector trays to "0." Remove trays.
3. Keep the projectors running on FAN until the lamps have cooled.
4. Switch the power to OFF on all units. Disconnect the dissolve unit from the projectors.

III. **Setting up the dissolve unit and operating it automatically or continuously (timer operation; not programmed ahead, no sound)**

A. Setting up

Follow the directions in InDepth II A, except for remote control and manual dissolve unit operations (6,8). Determine how to activate automatic continuous dissolve unit function. If there is no built-in timer, it may be possible to connect an external timer. (See Figure 12–2, Method B.)

Focus slides using FOCUS knob on each projector where option is available.

B. To show the presentation:
1. Set dissolve unit timer for desired intervals.
2. Turn the unit ON to activate the automatic timer.
3. To stop the automatic cycle, even temporarily, move the timer to MANUAL, or use stopping procedure appropriate to your dissolve unit.

C. After the presentation, follow the directions above, in InDepth II C.
D. To prepare slide trays for continuous operation, see Slide Projectors, InDepth IX B.

IV. Setting up the dissolve unit and operating it automatically with sound (programmed ahead of time)
   A. Setting up
      1. For slide projectors:
         Follow the directions in InDepth II A, except for remote control and manual dissolve unit operations (6,8).
         Focus slides using FOCUS knob on each projector, where option is available.
      2. For sound, see Figure 12–2, Method C. Use a tape recorder designed to synchronize slide/tape sound. Plug the tape recorder into the AC outlet. Attach the special cord from the tape recorder to the dissolve unit. (See Figure 12–3.) Insert the cassette tape side 1 into the tape recorder. If you are planning a continuous sound presentation where the program runs endlessly from beginning to end, obtain an endless loop cassette for the number of minutes of your presentation.
   B. To show the presentation:
      1. The dissolve unit switch should already be turned ON and the timer set to MANUAL if appropriate for your unit.
      2. Press PLAY on the tape recorder to start the presentation.
      3. To stop the presentation, even temporarily, press STOP on the tape recorder.
   C. After the presentation
      1. Press STOP on the tape recorder.
      2. Return the trays to "0." Remove the trays from the projectors.
      3. Keep the projectors running on FAN until the lamps have cooled.
      4. Switch the power to OFF on all units. Disconnect the dissolve unit from the tape recorder and from the projectors.

GLOSSARY

Black slide A blank slide with no content. Its purpose is to block out the white screen where no slide with visual content has been inserted. Slides that project black on the screen can be made by cutting unprocessed 35 mm film to fit a slide mount. You may use unprocessed expired film, if available.

Digital tone A particular type of pulse used with programmable dissolve units. It identifies which of several projectors is to be activated and what type of command the projector must perform.

Pulse A signal that activates visual and/or sound units.

Registration slides Slides with a simple pattern used to align the projectors with each other. They can be used as the first slides in each projector of a multi-projector program. They look like the "focus" frame of a filmstrip. You can buy them or shoot your own.

13 Filmstrip Projectors

CONTENTS

Operating Tips
- Front-Screen Units.............. 113
- Rear-Screen Built-In Units 113
- Sound Projectors 114
- Filmstrips...................... 114

Problems
- Visual Problems 114
- Filmstrip–Sound Synchronization Problems................... 116
- Sound Problems................. 117

InDepth
- Using the Projector 117
- Lenses for Front-Screen Projectors. 121
- Advancing Filmstrip/Sound Presentations............... 121
- Maintenance and Cleaning 122

Glossary...................... 122

Filmstrip projectors are widely used audio/visual machines. The cost of each projector is relatively low compared to that of other audio/visual equipment. These projectors are available in both front-screen and built-in rear-screen formats.

Front-screen projectors are useable for groups of all sizes, while built-in rear-screen equipment is primarily effective with small groups and individuals. Both formats are available as either silent or sound projectors.

OPERATING TIPS

I. **Front-screen units** (see Figures 13–2a and 13–3)

  A. Remove the projector from the case before operating, to allow better air circulation in the projector. Otherwise, you may damage the lamp, which will affect the clarity of your picture, or break the condenser lenses inside the projector.

  B. Remove loose papers from the desk or table on which you place the projector. Otherwise, the fan may suck them up, preventing proper air circulation or causing noise.

II. **Rear-screen built-in units** (see Figures 13–2b, 13–2c)

  Most of these projectors do not have cooling fans. Use the correct size lamp so that the unit does not overheat. After using the machine, allow it to cool before moving it.

114 *Filmstrip Projectors*

III. Sound projectors

Most machines have a jack labeled OUTPUT or HEADPHONE, with which to hook up either one headset or a jackbox to handle several headsets.

IV. Filmstrips

A. For a new filmstrip, prevent sticking by slowly running it through the projector two or three times.

B. Handle the filmstrip only by the edges. Do not allow it to touch the floor, as it will pick up dirt and grit that will scratch it when it is run through the projector.

C. The picture can be held on the screen for several minutes without damage to the filmstrip.

D. If you cannot improve the framing of the filmstrip, see if the framer was used properly. You may need to push or pull the framer while rotating it.

E. Consider recycling out-of-date filmstrips into slides. Cut the filmstrip frames you want and mount them in slide frames. The frames are commonly available at a camera shop. See also Slide Projectors, In-Depth VIII D.

PROBLEMS

PROBLEM	CAUSE	SOLUTION
Visual Problems		
GENERAL		
1. Black screen.	Lamp burned out.	Replace lamp.
	Cord not plugged in.	Check connections.
	Front-screen models contain both a fan switch and a lamp switch.	Put lamp switch ON.
2. First image on screen is two half-frames.	Filmstrip not framed.	Turn FRAMER knob (sometimes while pushing it in or pulling it out) until there is only one frame on screen.
3. Filmstrip does not feed properly, or sprocket holes appear on screen.	Filmstrip not correctly threaded.	Remove filmstrip and rethread, ensuring that holes are engaged in projector sprockets.
4. Projector does not engage filmstrip.	Filmstrip not pushed far enough into filmstrip slot.	Push filmstrip into slot until it can go no further. Turn ADVANCE knob.
FRONT-SCREEN PROJECTION		
5. Picture out of focus.	Lens barrel not properly adjusted.	Focus by turning lens barrel.

Filmstrip Projectors **115**

	Lens barrel not fitted in projector grooves.	Remove lens barrel from projector. Replace barrel, pushing it into grooves. When barrel clicks, turn it to focus picture.
	Where available, FOCUS knob not properly adjusted.	Adjust FOCUS knob.
	Film gate open.	Snap film gate shut.
	Filmstrip not properly seated in film path.	Open and close filmstrip gate a couple of times, or remove filmstrip and reinsert.
6. Lens barrel comes out of projector while you are focusing.	Projector too close to screen.	Replace lens barrel. Move projector away from screen so that barrel will focus while in projector.
7. Projected picture too small or too large.	Projector-to-screen distance needs adjustment.	If picture too small, move projector away from screen; if picture too large, move projector closer to screen.
8. Spots on screen.	Dirty projection lens.	Remove lens and clean both ends with lens tissue.
	Dirty condenser lenses.	Clean lenses with camel's hair brush.
9. Permanent line across screen.	Condenser lenses cracked.	Replace lenses or take in for repair.
10. Cover does not fit over case.	Machine not sufficiently or properly lowered into case.	Lower RAISE control on projector. For some cases, fit projector feet into cups in case.
11. Poor illumination.	Room not dark enough.	Darken room or bring projector closer to screen.
	Old lamp.	Replace lamp.
	Dirty projection lens.	Remove lens and clean both ends with lens tissue.
	Thin extension cord being used.	Use only heavy-duty cord.

116 *Filmstrip Projectors*

PROBLEM	CAUSE	SOLUTION
Visual Problems		
FRONT-SCREEN PROJECTION		
12. Filmstrip has bubbles in it.	Overheating due to incorrect lamp or missing heat filter, where available.	Check lamp code. Check that heat filter is in place, if used in your model.
13. Fuzz or hairs at edge of projected image.	Dirt in aperture and film gate.	Open film gate, clean gate and aperture.
REAR-SCREEN PROJECTION		
14. Unit overheats.	Incorrect lamp being used.	Check lamp code on projector nameplate.
15. Out of focus (especially older Singer machines).	FOCUS screw not properly adjusted.	Adjust screw until picture focuses. Use miniature screwdriver or coin. If threads are stripped, take in for repair.
16. Screen drops out.	Rough handling.	Take in for repair.
17. Filmstrip cannot be backed up to previous frame.	Some projectors ADVANCE only.	To reverse, use FRAMER knob. Push in or pull out while rotating, if necessary.
Filmstrip–Sound Synchronization Problems		
(For built-in combination unit or separate cassette unit designed to synchronize tape with filmstrip projector.)		
18. Filmstrip not advancing automatically.	Tape not programmed for automatic advance.	Advance filmstrip manually.
	Using "manual advance" side of tape.	Flip cassette over.
	Automatic advance signal on that tape will not work on your machine.	Advance filmstrip manually. Tape probably advances slide projectors.
	Defective machine.	Check tape on another machine; take machine in for repair.
	Dirty tape head.	Clean head.

Filmstrip Projectors 117

	Special patchcord needed for projectors used with separate cassette recorders.	Order equipment from dealer.
19. Sound and picture not synchronized (for additional synchronization problems, see Tape Recorders, Problem 18).	Tape and filmstrip not started simultaneously at correct place.	Rewind tape to beginning, and begin filmstrip on frame that says "start tape" or "sound here." Press PLAY on recorder.
	Tape rewound or advanced without adjusting filmstrip.	Locate filmstrip frame that coordinates with tape.
	Filmstrip advanced or reversed without adjusting tape.	Advance or rewind tape to match filmstrip frame.
	Filmstrip miscued in filmstrip–phonograph combination due to worn phonograph needle.	Replace phonograph needle.
	Dirty tape head.	Clean head.

Sound Problems

See Chapter 18 for any of the sound problems listed below. Unlike filmstrip–sound problems, these situations are not directly related to the functioning of the filmstrip. You may have sound problems when using either a self-contained filmstrip–sound unit or a separate tape recorder and filmstrip projector.

- No power.
- No sound on playback.
- Weak sound on playback.
- Fading in and out on playback.
- Mushy or distorted sound, wow and flutter, noise on playback.
- Humming or scratchy noise on playback.
- Incomplete erasure of previously recorded material.
- RECORD button does not depress.
- No RECORD button on machine.
- Tape sticks in machine; will not move.
- Tape squeals or squeaks.
- Tape runs backward on AC.
- Tape unthreads and jumbles while machine is running.

IN-DEPTH

I. **Using the projector (see Figure 13–1)**

A. Setting it up.

1. Remove projector from case. Plug into an AC outlet.
2. Insert the filmstrip (see In-Depth C).
3. Hold the film along the edges and push it into the filmstrip

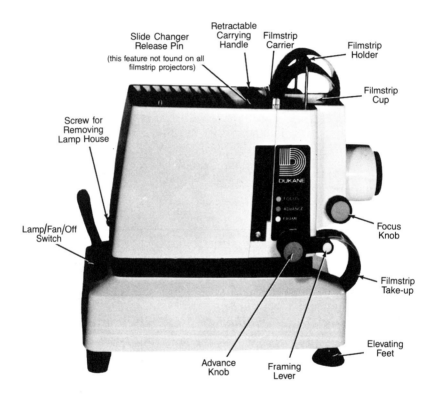

Figure 13-1 *Parts of a filmstrip projector. (Courtesy Dukane Corporation)*

slot until it can go no further. Now the film is ready to be engaged.

4. Turn the ADVANCE knob counterclockwise until the film comes out of the exit slot and is engaged by the sprocket wheel.
5. Turn the projector on. Turn the lamp on.
6. Focus the picture (see InDepth D).
7. Use the FRAMER if there are two half-frames on the screen when the filmstrip is advanced.
8. Where available, use the RAISE control to raise or lower the image on the screen. Turn the TILT control if the image is tilted on the screen, or place a book under one corner of the projector until the projector is level.
9. If using a sound projector, rewind the cassette tape to the beginning and advance the filmstrip to the frame that says "start the tape here." This will synchronize sound and picture.
10. Begin presentation.

B. After using the projector.
1. Remove the filmstrip.
2. Where possible, turn off the projection lamp.
3. Where available, leave the fan on to cool the lamp, until air from the projector is cool. Then turn it off.
4. Push the lens barrel back into the projector.
5. Put the projector back into the case and unplug.

C. Inserting the filmstrip (see Figure 13–2)

Front-screen: Insert the filmstrip upside down as you face the lens. The strip should curl toward the screen (for cartridge loading, see Figure 13–3).

Built-in rear-screen: The strip is fed differently, according to the particular projector. Try one of the following ways, holding the filmstrip as you face the projector (refer back to Figure 13–2):

1. If the filmstrip is to be pushed up into the slot, see Figure 13–2b.
2. If the filmstrip is to be pushed down into the slot, see Figure 13–2c.

The film is properly engaged if the ADVANCE knob pulls film through the projector smoothly.

D. Focusing.
1. *Front-screen projector*: The projector is focused by moving the projection lens barrel back and forth. Turn either the FOCUS knob or the lens barrel itself. Best focus will be achieved by moving the lens past the point of focus and then back to the focus point.
2. *Rear-screen projector*: Have on hand small screwdrivers, Phillips or flat. On most of these

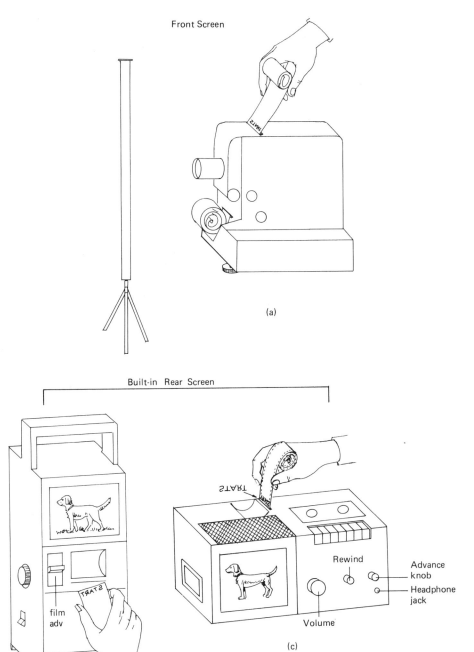

Figure 13-2 *How to insert filmstrip.*

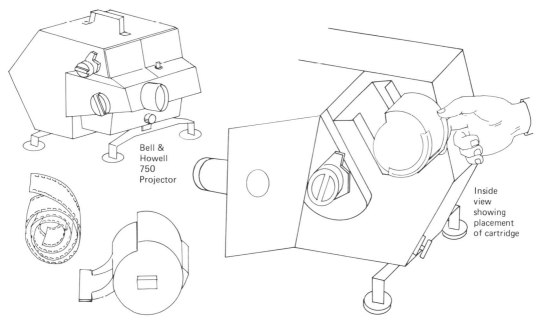

Figure 13-3 *How to load filmstrip in a cartridge. Though this Bell & Howell model is no longer made, many units are still being used.*

machines, the focus is preset at the factory. However, there is sometimes a FOCUS knob for more precise adjustment. Some equipment has a focus screw on top or on the back of the machine, which will refine focus. Otherwise, take equipment in for repair.

II. **Lenses for front-screen projectors**

There is a variety of projection lenses available, according to the size of the viewing audience. Unless you specify otherwise when ordering a particular brand and model, you will get a particular size lens. Auditorium lenses cost three to four times the price of classroom lenses. Classroom lenses are adequate for many auditorium conditions. Try out the projector in the location where it will be used most often to determine your lens needs.

III. **Advancing filmstrip/sound presentations**

A. General information

Filmstrip/sound presentations are designed to synchronize or coordinate picture advance with the audio. The presentation comes as a filmstrip and cassette, or a filmstrip and record.

Filmstrip/sound presentations can be synchronized automatically by the projector or manually by the projectionist. There are two ways to automatically advance filmstrip/sound programs. First, use a projector with built-in cassette or phonograph. This unit is actually two machines built into one. Second, you may use certain silent filmstrip projector models with a special cassette unit designed to synchronize 50 hz signals. These two machines must be connected with an appropriate cord. Consult your audio/visual dealer. For manual advance, you may use a silent projector and a separate average cassette unit, with the two machines not connected to each other.

B. Automatic advance with tapes
Tapes programmed to automatically advance filmstrips have inaudible 50 hz signals. The signals must be inserted with special equipment. Commercial filmstrip/cassette presentations are standardized at 50 hz. They are therefore compatible with, and the signals can be activated by, filmstrip/sound synchronizing units.

C. Manual Advance
1. Playback: In addition to inaudible automatic signals on one side, commercially-made filmstrip/sound programs may contain the same program with audible signals for manual advance on the other side of the cassette or record. If the tape or record does not have audible signals, follow a script during the performance.
2. Recording: When developing your own sound, if you do not have a special tape recorder that emits a pulse, you can still coordinate pictures and sound. At appropriate intervals while recording the program, establish your own audible signals on tape by using a clicker or clinking a glass.

IV. Maintenance and cleaning
A. Front-screen projectors:
1. Condenser lenses need to be cleaned, but are not always easily removable. Reach in with a camel's-hair brush to dust them off.
2. Projection lenses need to be cleaned regularly. The lens barrel is usually removable. Clean both the front and back ends of the lens with lens tissue.
   Glass pressure plates that hold the filmstrip in place need to be cleaned or dirt on them may scratch the filmstrip.
3. For sound projectors using cassettes, use head-cleaning cassette tape after forty hours of playing time. Instructions accompany the cleaning cassette. See also, InDepth IX, Tape Recorders (Cassettes).
4. Use a wooden toothpick to gently clean dirt from the film channel.
B. Built-in rear-screen projectors:
Clean the screen with art gum eraser, or wash with soft cloth moistened with very mild soap solution. Rinse off soap, and dry surface.

GLOSSARY

Aperture Opening in the film gate. Light from the lamp passes through this opening and hits the film.

Condenser lens Lens that concentrates light through the image to be projected. Located inside the projector.

Film gate Device that holds the film in place within the projector.

Heat filter Special filter lens that prevents the heat of the projector lamp from burning the filmstrip.

Lens barrel See Projector lens.

Projector lens Lens that focuses and projects image on the screen. Also called the *lens barrel* on front-screen projectors.

Pulse A signal on tape or record that determines when the filmstrip is to be advanced or stopped. It can be audible or inaudible.

Sprocket holes The holes on either side of the filmstrip that are engaged by the sprocket wheel, so that the filmstrip will feed evenly into the projector.

Sprocket wheel The metal or plastic wheel with teeth that engages the filmstrip in the projector. Also called sprockets.

14 Movie Cameras

CONTENTS

Operating Tips
- Camera 124
- Shooting and Editing 125
- Film 125
- Battery 125
- Sound 125

Problems
- Shooting 126
- Camera 126
- Sound 127
- Processed Film
- Film 127

- Sound 129

InDepth
- Operating Procedures 129
- Removing Partially Exposed
  Super 8 Cartridge 131
- Using Filters 131
- Types of Film 131
- Running Times of Film 132
- Making Sound Movies without a
  Sound Movie Camera 132
- Cleaning Camera 133

Glossary 133

Moviemaking can be a very enjoyable experience. The purpose of using a movie camera instead of a still camera is to tell a story through some action. Movie cameras come with various features to enhance your filmmaking project, but you can still have a good production with a basic camera. Most cameras come equipped with some of the following features: sound capability, single-framing (to make animations), fast- or slow-motion speeds, camera-mounted floodlight (to provide extra light), low-light filming capability, and zoom lens. Some models allow interchangeable lenses. The price of the camera depends on its features and accessories.

Super 8 movie cameras are rapidly being replaced by video camera equipment.

The information in this chapter is designed primarily for the super 8 user. However, the operating tips, problems, and glossary can apply to both super 8 and 16 mm cameras.

OPERATING TIPS

I. Camera tips

   A. Check the batteries to see if they are good.

   B. For temperatures below freezing,

protect the camera inside your coat or car until you're ready to shoot.

C. When a cartridge is in place, shut the film door securely before operating the camera.

D. Hold the camera steadily when shooting. Use a tripod whenever possible.

E. Depress the camera trigger fully.

II. **Shooting and editing tips**

A. Plan your story before you shoot. Use the storyboard technique (see Glossary).

B. Start shooting slightly before the action begins and for a short time after it ends.

C. Keep a written record of your shots and/or scenes.

D. When shooting several scenes, vary their length, pace, and subject-to-camera distance.

E. If you plan your scenes carefully, you will minimize the need for later editing. Film editing is simply a cut-and-paste job; you cut out the unwanted segments and splice together the desirable ones. The most basic equipment consists of a silent or sound editor to view the film and a splicer to cut and reassemble the segments. Instructions accompany the equipment.

III. **Film tips**

A. Most sound cameras accept silent film; however, silent cameras will not accept sound film.

B. Know how the film counter or indicator on your camera works, particularly when you need to remove a partially used cartridge.

Needle begins at   Goes to
     50 —————————— 0
Needle indicates how much film is left.

Needle begins at   Goes to
      0 —————————— 50
Needle indicates how much film has been exposed.

C. Usually insert the film cartridge with the labeled side up. Snap it into the camera.

IV. **Battery tips**

A. Silent movie cameras may have both motor batteries and a separate light meter battery; sound movie cameras sometimes have an additional sound battery.

B. Various types of batteries are located in different areas of the camera body.

V. **Sound tips**

A. Check the sound input during filming by monitoring the camera with an earphone.

B. If you're using a separate, rather than built-in microphone, place the microphone at least three feet from the camera to minimize camera motor sounds.

C. Move the microphone as little as possible during recording.

D. Indoors, try to shoot in a carpeted room. Outdoors, beware of traffic and wind sounds, and use a windsock for a microphone not built into the camera (see Glossary).

126 *Movie Cameras*

PROBLEMS

PROBLEM	CAUSE	SOLUTION
Shooting Problems		
CAMERA PROBLEMS		
1. No picture through viewfinder.	Lens cap on.	Remove lens cap.
	Built-in lens cover over lens.	Flip cover up.
2. Camera will not operate properly.	Batteries old.	Check. Replace.
	Batteries inserted incorrectly.	Check positive and negative (+ −) terminals in battery compartment.
	Battery and/or camera contacts dirty.	Clean battery contacts with rough cloth. Clean camera contacts with a pencil eraser.
	Battery cover and/or film door open.	Shut door(s).
	Trigger not fully depressed.	Depress trigger fully.
	Defective camera.	Take in for repair.
3. Low light signal appears.	Not enough light.	Wait for more light, or add artificial light.
	If low light signal appears in bright light, light meter is obstructed.	Remove hand or other object covering light meter.
	Light meter battery either old, dirty, or incorrectly inserted.	Check light meter battery, if there is one.
4. Low light signal flickers on and off.	Amount of light questionable.	Take a chance filming, but movies may be unuseable. It is safer to add artificial light.
5. Motor does not run.	Motor batteries old.	Check. Replace.
	Motor battery and/or camera contacts dirty.	Clean battery contacts with rough cloth. Clean camera contacts with a pencil eraser.

Movie Cameras 127

	Motor batteries inserted incorrectly.	Check positive and negative (+ −) terminals in battery compartment for each battery.
6. Trigger sticks.	RUN–LOCK button not fully released.	Release RUN–LOCK button.
7. Door of camera will not shut.	Film cartridge inserted improperly.	Remove cartridge; reinsert. You may need to turn it around.
8. Motor sound not normal.	Chattering sound: Film running through improperly.	STOP FILMING. Examine cartridge. It may need to be replaced. If sprocket holes are torn, advance film manually past torn point.
	Slowed-up sound: Motor batteries dirty or worn.	Clean or replace motor batteries.
	Cold motor batteries.	Keep batteries at room temperature (warm with your hands for ten to twenty minutes).

SOUND PROBLEMS, USING SOUND CAMERA

9. No light or needle indicator in camera viewfinder to show sound recording being made, or no sound when you are monitoring recording through an earphone.	Using silent cartridge film.	Use sound cartridge.
	Remote microphone not plugged in.	Check microphone connection.
	Faulty remote microphone cable.	Replace microphone.
	Sound batteries in camera old.	Check. Replace sound batteries.
	Trigger not fully depressed.	Depress trigger fully.
	No sound in scene.	See InDepth VI.
	Motor running but film not moving.	Try another cartridge.

Processed Film Problems
FILM PROBLEMS

10. Processed film black (unexposed).	Film shot with lens cap on.	Reshoot segment.

128 *Movie Cameras*

PROBLEM	CAUSE	SOLUTION
Processed Film Problems		
FILM PROBLEMS		
	Defective camera.	Take camera in for repair.
	Defective film cartridge.	Replace cartridge.
11. Clear film or yellow-orange edges.	Accidental exposure to light.	Reshoot segment.
	Light leak in camera.	Take camera in for repair.
12. Poor color.	Film processed after expiration date.	Shoot again, and process film before expiration date.
	Incorrect film or filter setting used for lighting conditions.	Check filter and use appropriate film.
13. Scene appears too light.	Light meter batteries cold, if camera has light meter batteries.	Use batteries at room temperature. Keep camera at room temperature.
	Automatic exposure setting not switched ON, or manual f stop setting is incorrect for scene.	Check exposure setting.
14. Scene appears too dark.	Bright light shining directly into light meter.	Balance lighting. Reshoot segment.
	Not enough light.	Add floodlights or move your shooting location.
	Automatic exposure setting not switched ON, or manual f stop setting is incorrect for scene.	Check exposure setting.
15. Pictures blurred.	Camera not focused, if there is a focusing ring.	Focus carefully, where possible.
	Camera not held steadily while shooting.	Use tripod to hold camera steady, or brace your body, as follows:
		Stand against a wall, where possible.
		Place elbows as closely as possible to sides of body.
		Place one leg slightly forward, one slightly back of body.

Movie Cameras

	Camera panned across scene too quickly.	Pan shots should rarely be used (see Glossary).
16. Subject moves too fast on screen (like old-time movies).	Camera speed at fewer frames per second than normal, such as 9 or 12 fps. The fewer fps, the faster the action.	Set speed to 18 fps for super 8 silent or sound film. Some models operate at 24 fps.
	Cold motor batteries.	Keep camera at room temperature (warm batteries in hands for ten to twenty minutes).
17. Subject moves too slowly on screen (unintended slow-motion).	Camera speed at more fps than normal, such as 36 or 54 fps. The more fps, the slower the action.	Set speed to 18 fps for super 8 silent or sound film. Some models operate at 24 fps.
	Camera SLOW-MOTION switch used.	Flip SLOW-MOTION switch to OFF.

SOUND PROBLEMS

18. Sound distorted when exposed film projected.	Dirty sound mechanism in camera.	Use special cleaning materials; consult your dealer.
		Reshoot segment.

See also Problem 12, Super 8 Movie Projectors.

IN-DEPTH

I. Operating procedures

(See Figure 14–1.)

A. Select appropriate film (see In-Depth IV).
B. Check the motor batteries to see if they are good.
C. Set the camera speed for the results you want, if there is a variable-speed feature (see Table 14–1).
D. Adjust the filter switch for the film you are using (see InDepth III.)
E. Where possible, adjust the viewfinder eyepiece at the back of the camera to suit your vision. When using a zoom lens, look at a faraway object, turn the zoom to its greatest closeup, and focus the lens on infinity. Then adjust the eyepiece until the object appears

Table 14–1

Desired Action When Film Projected at Normal Speed	Super 8 Camera Running Speed (Silent or Sound)
Normal	18 fps (some models are 24 fps)
Slow-Motion	24 fps or more
Fast (old movies)	12 fps or less

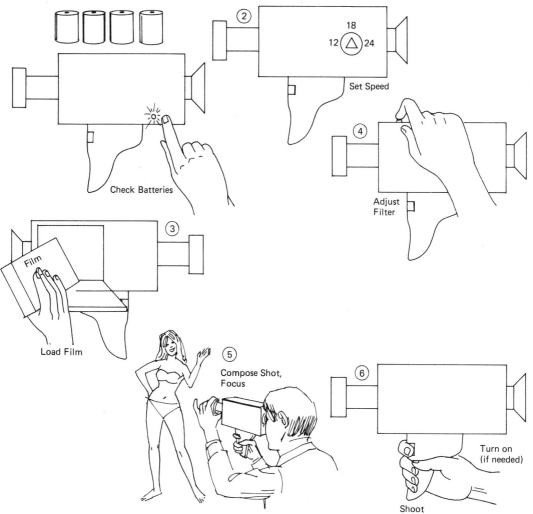

Figure 14-1 *Using your movie camera is easy.*

sharp to you. The eyepiece only needs to be adjusted for another user.

F. Load the film, open film door, insert cartridge, close door, run film to 0 or 50, depending on your camera (see Operating Tips IIIB).

G. Steady the camera or set the camera on a tripod.

H. Compose your shot.

I. Focus the camera: estimate subject-to-camera distance, and adjust the focusing ring accordingly.

If you are planning to zoom in and out with a zoom lens, rotate the zoom control to the greatest closeup position. Focus. Adjust the lens to the desired position for the initial shot.

J. Where available, turn ON/OFF switch to ON.

K. Press the trigger to shoot the film.

II. Removing a partially exposed super 8 cartridge

About 4 inches of used film will be ruined—more if it is sound film. You will not lose exposed film if you run the camera for one or two seconds beyond the last scene before removing the cartridge. But you will then have a blank space when the film is projected, which you may want to splice out.

If you must remove the cartridge:

A. Note the amount of film left on the film indicator before removing the cartridge, and mark it on the cartridge when taking it out of the camera.

B. When you replace the cartridge, the film indicator will return to the beginning. Therefore, you must determine for yourself how much film is being used as you are shooting.

III. Using filters

A built-in filter is used to adapt the camera lens for film use indoors or outdoors. You cannot see it by looking into the camera lens. (See InDepth IV for use of filter with particular films.)

Depending on your camera, control the filter in one of two ways:

1. ON/OFF switch.
2. Filter key. Turning the key in may either remove or replace the filter; this depends on your particular camera. Consult your owner's manual or a dealer. If the key has been lost, a new one may be ordered from your dealer.

Add-on filters are available to correct lighting conditions or to provide special effects. You may be able to mount your still-camera filters using filter adapting rings.

IV. Types of film

Most super 8 films are in color. The films listed in Table 14–2 are made by Kodak.

Table 14–2

Sound/Silent	Film Type	Lighting Conditions	Filter*
Silent (or Sound)	Kodachrome 40, Type A	Outdoors	ON
		Indoors, use tungsten light or movie floods.	OFF
Sound	**Ektachrome 160 Type A	Daylight or fluorescent.	ON
		Tungsten lights	OFF
Silent	**Ektachrome 160 Type G	For all types of lighting including fluorescent and low light.	OFF

*When the scene is illuminated by more than one type of light and you are not sure whether to use the filter, keep it turned on.

**If you have an older camera, check with your dealer to see if it will handle this film.

V. Running times of film

It takes three minutes and twenty seconds to shoot and project one roll of super 8 film (50 feet) when shot at 18 fps. You may find it useful to time your segments, as shown below in Table 14–3.

VI. Making sound movies without a sound movie camera

A. Tape record the sound separately while you are filming with a silent camera. Play back the tape when the film is being projected on the screen.

Table 14–3

Running Time of Film for Common Projection Speeds*				
Film format	Super 8 (72 frames per foot)			
Projection speed in frames per second	18 (normal)		24** (slow motion)	
Running time and film length	feet	frames	feet	frames
Seconds 1	0	18	0	24
2	0	36	0	48
3	0	54	1	0
4	1	0	1	24
5	1	18	1	48
6	1	36	2	0
7	1	54	2	24
8	2	0	2	48
9	2	18	3	0
10	2	36	3	24
20	5	0	6	48
30	7	36	10	0
40	10	0	13	24
50	12	36	16	48
Minutes 1	15	0	20	0
2	30	0	40	0
3	45	0	60	0
4	60	0	80	0
5	75	0	100	0
6	90	0	120	0
7	105	0	140	0
8	120	0	160	0
9	135	0	180	0
10	150	0	200	0

*Kodak, *Movies with a Purpose*, p. 13.
**With some projectors, 24 fps is normal speed.

B. Record the sound on silent film after it has been processed: First, have a magnetic stripe added to your film. Then, a special projector will allow you to record the sound and play it back. You can erase and rerecord information.

C. For other options, consult your dealer.

VII. Cleaning the camera

A. *Lens.* Do not use eyeglass tissue or facial tissue. Breathe on the lens and wipe gently with special camera lens tissue or clean, soft, dry cloth. For stubborn stains, use one drop of lens cleaning fluid on lens tissue, *not* on the lens. Wipe the lens gently.

B. *Interior of camera.* Clean occasionally with canned air and a camel's-hair brush. Spray air on your hand once before using on the camera. Use cotton swabs dipped in alcohol to clean the aperture plate.

C. *Batteries.* Clean the terminals on the battery with rough cloth. Clean camera battery terminals with a typewriter or pencil eraser.

D. *Sound recording mechanism, where available.* If the sound is distorted on a processed roll of film, use special sound-cleaning materials on the camera before shooting the next roll. See your dealer.

GLOSSARY

Animation Movie technique used to bring to life or provide motion for inanimate objects. A camera needs to be equipped with single-frame capability. This enables the movie camera to shoot one frame at a time, as with a still camera.

ASA American Standards Association. A standard of measuring film speed or film sensitivity to light commonly used until 1983. Now film speeds are labeled ISO. See ISO.

Available light Also called existing light; for example, daylight, table lamps, church interiors, fluorescents, candlelight, Christmas lights, and so on.

Canned air Can of compressed air, available at a photo dealer. Used for cleaning delicate parts of a camera or lens that should not be touched too often, if at all.

Cut A sudden change from one shot to the next.

Editing Process of collecting, arranging, and splicing various film segments into a finished movie.

Emulsion The light-sensitive coating on the film, on which the image is registered.

Establishing shot Overview, rather than close-up, picture. Usually used at the beginning or end of a filming sequence to identify subject, location, or action to follow.

Fade Technique in which there is a gradual lightening of the picture at the beginning of a scene or a gradual darkening of the picture to black at the end of a scene. Some cameras have a fade-in/fade-out button to achieve this effect. Effect is similar to that of a rheostat you might use in your dining room.

Fast motion Projected film shows the subject moving fast in stilted fashion, as in old-time movies. To achieve this effect with a super 8 camera, the subject behaves normally during filming, but the camera is set at 12 fps, or less, where this option is provided.

Film indicator Device on camera to tell you how much film is left or how much has been exposed (see Operating Tips III B).

fps Frames per second. The number of still pictures or frames a movie camera photographs in one second.

ISO International Standards Organization. The standard for measuring film speed. The higher the number, the faster and more sensitive the film is to light. Most modern movie cameras are made to adapt to varying film speeds. Therefore, no adjustments are necessary when film is inserted into the camera. A notch in the cartridge sets the ISO (see also ASA).

Pan Shot taken while moving the camera from left to right, or vice versa. Do this sparingly. Pan very slowly. If you must pan, use a tripod and test your shot ahead of time. Camera should hold a still shot at beginning and end, and camera movement in-between should be imperceptible.

Slow motion Projected film shows the subject moving slowly in a floating fashion. To achieve this effect with a super 8 camera, the subject behaves normally during filming, but the camera is set at 24 fps, or more, where this option is provided.

Splicing Process of joining together two pieces of film.

Splicer Simple device to cut and/or join together two pieces of film. Dry splicers use splicing tape. Wet splicers use a special cement. Instructions accompany the equipment.

Splicing tape Thin, clear tape used with a splicer to join or repair film. Usually has the sprocket holes to match your film sprocket holes (with super 8 film, use super 8 splicing tape, and so on).

Storyboard Paper and pencil technique used to organize and coordinate ideas for visuals and script in sequential order. Looks similar to a comic strip sequence.

Trucking shot Filming technique in which the camera moves alongside a subject.

Windsock Inexpensive foam-rubber jacket used to cover a microphone. Available for most microphones that are not built in. Useful in eliminating whistling wind noise.

Wipe Technique used to visually join two scenes. At the end of one scene, a black card is brought down in front of the lens, covering it so the picture is blacked out. The camera is then stopped. The next scene is begun with the lens still covered by the card. After the camera has been started, the card is lifted out of the way. It is as if you were opening or closing an opaque window to start or end a scene.

Zoom Camera effect achieved through the use of a special zoom lens that allows close-up (telephoto) and wide-angle shots without moving the camera.

15 Super 8 Loop Projectors

CONTENTS

Operating Tips
    Projectors 135
    Cartridges 135
Problems 137
InDepth 137

Although super 8 loop projectors are losing their popularity, many are still in use. Film for the super 8 loop projector is contained in a plastic cartridge; you never touch the film itself. You can stop the film at the end of your presentation or allow it to continue indefinitely. No rewinding or threading is ever necessary, as the film is an endless loop. These projectors usually do not function in reverse.

OPERATING TIPS

I. Projectors

(See Figure 15–1.)

Projectors have been available in a variety of formats:

1. Silent or sound.
2. Large-screen.
3. Small 8 inch by 10 inch built-in screen.
4. Built-in screen, convertible to large-screen.
5. Loop projector, convertible to standard reel-to-reel projector.

II. Cartridges

Cartridges are not interchangeable. Each manufacturer puts out a cartridge to fit its own projectors.

Silent cartridges usually can handle up to fifteen minutes of film, and sound cartridges up to thirty minutes.

A sound cartridge containing silent film could be played on a sound projector.

Empty cartridges are sold so that you can make your own film loop, or you can have your film put into a loop cartridge when it is processed.

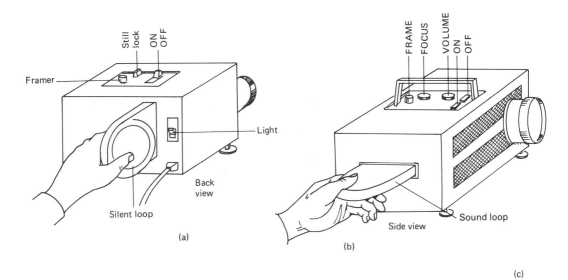

Figure 15-1 (a), (b) Front-screen loop projectors. (c) Rear-screen loop projector. (Courtesy Dumont Instrumentation, Inc.)

PROBLEMS

PROBLEM	CAUSE	SOLUTION
1. Film does not start.	Film not properly engaged.	Pull out and reinsert cartridge a few times. Peek through cartridge to inspect film for damage.
2. Film stops.	Film damage, such as torn sprocket holes.	Do not splice bad film: With your finger, move damaged film past open part of cartridge and reinsert cartridge in projector, or send cartridge back to manufacturer.
	At end of loop.	Push START button again.

IN-DEPTH

Procedure for using projector:

1. Push the cartridge into the film slot.
2. Turn the projector on. For sound projectors, adjust the VOLUME control.
3. Focus the lens by turning the lens barrel or by adjusting the FOCUS knob.
4. Adjust the FRAMER, when necessary, so there is one complete picture on the screen rather than halves of two pictures.
5. Use the STILL FRAME control, where available, to stop and hold the picture on the screen momentarily.

16 Super 8 Movie Projectors

CONTENTS

Operating Tips 138
Problems
    General Problems 139
    Picture Problems 140
    Sound Problems 141
    Film Problems 141
InDepth 143

The super 8 movie projector is used primarily for homemade or classroom-made movies rather than commercial films. However, since old films can be transferred to the video format, it is now being rapidly replaced by video equipment.

Most super 8 projectors have the reel-to-reel format. A few projectors that are self-threading use either cartridge or reel-to-reel formats. You may purchase a cartridge to house the reel-to-reel film for this type of projector.

There are various features available:

1. Silent.
2. Sound, some with recording capability.
3. Still framing.
4. Slow-motion projection.
5. Automatic threading (on most projectors).

What super 8 film looks like:

1. Super 8 film has a set of holes (sprocket holes) running down only one side of film.
2. On silent film, the non-holed side has a black line.
3. On sound film, the non-holed side usually has a brown magnetic stripe, like audio tape. (There may also be an additional brown stripe on the sprocket hole side.) This is called a magnetic sound track. Some projectors have been made with both optical and magnetic sound capability (see Glossary, 16mm Projectors).

OPERATING TIPS

1. ALWAYS stay in the room while the projector is in operation.
2. Do not use super 8 film in a projector

Super 8 Movie Projectors **139**

equipped for 8mm only, or vice versa; the size of the film sprockets is different. Today, however, most super 8 projectors can handle both super 8 and 8mm film at the flip of a switch.

3. Super 8 silent film can be run through a super 8 sound projector (and vice versa). Action may be slightly faster on the screen.

4. The empty take-up reel should be at least as large as the full reel to take up all the film.

5. The take-up reel should not pinch the film; otherwise, the film will not flow smoothly through the projector.

6. Handle the film by the edges only.

7. Do not let the film touch the floor.

8. The lens and the lamp are accessible on all projectors, even though it may take some thinking about how to reach them.

9. Stop the projector before using the REVERSE switch.

10. Light output is reduced when the film is on STILL. A focus adjustment may be necessary.

11. When "END" appears on the film, turn the projector switch to MOTOR until the lamp cools. Then turn the switch to OFF. If there is no motor switch, turn the projector off but do not move it until it cools.

12. Do not wind a dirty power cord around the take-up reel.

13. Be prepared if the film breaks! Buy an inexpensive splicer, and splicing tape or cement. Simple directions for joining two pieces of film or repairing film come with the splicer (see Glossary, Movie Cameras).

PROBLEMS

PROBLEM	CAUSE	SOLUTION
General Problems		
1. Projector does not operate.	Power not ON.	Turn projector switch to ON.
	Power cord not well connected.	Check connection; check plug adaptor, if there is one.
	No current.	Test outlet, using other equipment.
2. Motor and sound, but no picture on screen.	Lamp burned out.	Replace lamp.
	Lamp switch not on.	Turn lamp switch to ON.
3. Picture looks too narrow at top or bottom (keystoning).	Projector tilted too much in relation to screen.	Set projector lens perpendicular to screen by adjusting projector or screen height and angle (see Overheads, InDepth IB, and Figure 3–2.)

140 *Super 8 Movie Projectors*

PROBLEM	CAUSE	SOLUTION
Picture Problems		
4. Poor illumination.	Room not dark enough.	Darken room further, or bring projector closer to screen.
	Wrong lamp.	Check specified lamp code on projector nameplate (see also Lamps).
	Projection lamp not properly inserted.	Check lamp placement.
	Old, dark-looking lamp.	Replace lamp.
	Greasy, dirty projection lens.	Clean lens.
	STILL setting being used.	STILL setting usually cuts down on lighting.
5. Two half-pictures on screen.	Film not framed.	Turn FRAMER knob until there is only one picture on screen.
6. Cannot focus.	FOCUS knob will not work because lens barrel off gears or gears are loose.	Push lens back into lens hole. Gears will engage. Use FOCUS knob. If loose gears, move lens back and forth by hand.
	Cracked projection lens.	Replace projection lens.
7. Fuzzy picture.	Greasy, dirty projection lens.	Remove lens. Wipe both ends of lens with lens tissue.
	Picture out of focus.	Use FOCUS knob.
	Cracked projection lens.	Replace projection lens.
	Condensation on lens when projector is brought in from cold.	Wipe lens very gently with clean facial tissue. Let projector warm up twenty to thirty minutes.
	Old, brittle film.	Cannot repair.
8. Picture frequently in and out of focus.	Film gate/aperture problem.	Take in for repair.
9. Constant spots in one area of screen.	Greasy, dirty projection lens.	Remove lens; wipe both ends of lens with lens tissue.

10. Fuzz projecting into picture.	Dirt in aperture.	Stop projector. Lightly clean aperture, using lens brush supplied with projector or cotton swab and alcohol.

Sound Problems

11. No sound.	If using detachable speaker, may be cord or connector trouble.	Check cord or connector.
	Wrong projector for film sound track.	Check your film. A fine wiggly line down one side requires an optical sound projector; a solid brown stripe requires a magnetic sound projector.
	Silent film being used.	If film does not have brown stripe opposite sprocket holes, it is silent film (see Movie Cameras, InDepth VI).
	Sound switch not ON.	Turn sound switch to ON.
	Volume low.	Raise volume.
	Silent projector being used.	Locate sound projector.
12. Garbled, fuzzy sound, low volume.	Dirty sound head.	Clean head with cotton swab and alcohol.
	Old film.	Cannot repair.
	Film not passing across sound heads properly.	Stop film. Check path of film.

Film Problems

13. Film does not automatically thread.	STOP PROJECTOR IMMEDIATELY.	Set motor switch at STOP, then REVERSE, so film will exit at starting point. Determine cause as listed below. Reinsert film.
	Film not engaged by projector.	Manually feed film into machine as far as it will go. Then turn to FORWARD or THREADING.

PROBLEM	CAUSE	SOLUTION
Film Problems		
	Projector not in threading position, where available.	Set projector on THREADING.
	No visible reason.	Use REVERSE, then reinsert film.
	Projector designed for manual threading only.	Thread film by hand.
	Motor switch not on FORWARD.	Place motor on FORWARD.
	Film leader not trimmed.	Trim with scissors or built-in trimmer.
	Film leader damaged.	Cut damaged portion. Trim remaining leader with built-in trimmer.
	Lens housing not closed.	Close lens housing.
	Film path has film trimmings.	Clean film path with toothbrush, then soft cloth dampened with alcohol.
14. Projector makes chattering, clicking noise; film flickers.	STOP PROJECTOR IMMEDIATELY.	
	Upper or lower loop too small.	(1) Press loop restorer and turn projector to FORWARD; or (2) Keeping projector off, disengage film and allow a couple of extra frames to make a larger top loop; or (3) rethread film.
	Lens housing not fully closed.	Close lens housing.
	Film gate coated with residue from new films.	With alcohol and cotton swab, clean aperture plate until shiny.
	Sprocket holes chewed up.	Cut bad part of film; splice film with splicer using splicing tape or cement and rethread projector. If film borrowed, put flag in film reel at damaged section.

		Check film gate; with alcohol and cotton swab, clean it until shiny.
15. Film scratched.	STOP PROJECTOR IMMEDIATELY.	See if problem is in projector or if film is already scratched.
	Dirt or emulsion accumulated in film path.	Clean film path with toothbrush, then soft cloth dampened with alcohol.
	Film itself scratched.	No cure.
	Damaged parts on projector.	Take in for repair.
16. Film breaks; film jams, and piles up in projector.	STOP PROJECTOR IMMEDIATELY.	Remove film with tweezers if necessary. Follow procedures in Projectors, InDepth IIIC, 16mm.
	Sprocket holes chewed up.	Cut bad film; splice film, rethread projector. If film borrowed, put flag in film reel at damaged portion. Save bad film and return to lender with reel.
	Poor splice.	Resplice film.
	Film reels not snapped onto projector.	Repair film. Snap reels onto projector.

IN-DEPTH

Three areas of the projector need cleaning:

1. *Film path.* Clean the film path with the brush that is provided or a toothbrush. Then use a soft, clean cloth dampened with alcohol. Clean the aperture area and film gate regularly, using cotton swabs dipped in alcohol, until they are shiny. Use a wooden toothpick for small corners.

2. *Lens.* Projector should be cooled off before cleaning the lens. The projection lens should be cleaned regularly on both ends with lens tissue or a soft, dry cloth.

3. *Sound area.* The sound head is similar to a tape recorder head. It needs cleaning with cotton swabs and alcohol or cassette head-cleaning fluid (see also 16mm Projectors, InDepth VI C).

GLOSSARY

See next chapter.

17 16mm Movie Projectors

CONTENTS

Operating Tips
- General 144
- Light and Lamps 145

Problems
- General 145
- Picture 146
- Sound 149
- Film 151

InDepth
- Parts of a Projector 155
- Threading Slot-Load Projectors 156
- Threading Autoload Projectors 157
- Threading Manual Projectors 161
- Rewinding Film 163
- Cleaning Your Projector 163
- Running Times for Various Size Reels 163

Glossary 164

A motion picture is actually a series of still pictures passed through a projector so quickly that the eye perceives movement. The most common form of motion picture used by organizations and schools is 16mm. There are both sound and silent film formats, the latter distinguishable by the sprocket holes on both sides of the film.

Every projector of sound films has three elements:

1. Two reels, one to feed film into the machine and the other to take up the film.

2. Several sprockets and rollers to guide the film through the projector.

3. A sound system to decode the sound on the passing film, turn it into a small electric current, amplify the current, and turn it back into sound through a loudspeaker.

Projectors cost from several hundred dollars to more than $1000.

OPERATING TIPS

General tips

1. Threading diagram for your machine is usually on the inside of the cover. When threading a manual or autoload machine, make sure the machine is in the threading position. Make sure fully on the sound drum. See Figure 17–1.

2. On many projectors there is a silent/sound lever for film with or without sound.

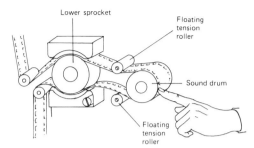

Figure 17-1 *Make sure film is fully on sound drum.*

Change the lever when the projector is in FORWARD. (To deal with garbled sound, see Figure 17-2 and Problem 18.)

3. The empty take-up reel should be at least as large as the full reel to take up all of the film.

4. The take-up reel should not pinch the film; otherwise, the film will not flow smoothly through the projector.

5. Check the film gate periodically, especially after using new films. If the film gate is not shiny and there is a sticky residue, use alcohol and cotton swabs to clean. Use wooden toothpicks for stubborn deposits.

6. Handle the film by the edges only.

7. Do not let the film touch the floor.

8. REWIND and REVERSE are two different functions (see Glossary.) Always stop the projector before using the REVERSE switch.

9. Do not wind a dirty power cord around the take-up reel.

10. Be prepared if the film breaks! Buy an inexpensive splicer, splicing tape or cement, and become familiar with how to use it. Simple directions for joining or repairing film come with the splicer (see also Glossary, Movie Cameras).

Light and Lamps

1. The projector lamp should be on NORMAL, not HIGH. The life of the lamp increases approximately ten times.

2. Exciter lamps used in the sound system last much longer than projection lamps (see Glossary, Optical sound system).

3. Light output is reduced when the film is on STILL. Focus adjustment may be necessary.

4. When "END" appears on the film, turn the switch to FAN or MOTOR until the lamp cools. Let the lamp cool for five to seven minutes. Then turn the switch to OFF.

PROBLEMS

Problems arise mostly during the first two and one-half minutes of a showing or at the end. Make only minor repairs, those on the front of the projector.

PROBLEM	MANUAL (M) OR AUTOLOAD (A)	CAUSE	SOLUTION
General Problems			
1. Projector does not operate.	M, A	Power not ON.	Turn projector switch to ON.
		Power cord not well connected.	Check connection; check plug adaptor.

146 16mm Movie Projectors

PROBLEM	MANUAL (M) OR AUTOLOAD (A)	CAUSE	SOLUTION
General Problems			
		No current.	Test wall outlet, using other equipment.
2. Motor and sound but no picture on screen.	M, A	Lamp burned out.	Replace lamp.
		Lamp switch not ON.	Turn lamp ON.
		Motor switch on FORWARD but not on PROJECT.	Turn switch to PROJECT.
3. Picture looks too narrow at top or bottom (keystoning).	M, A	Projector height tilted too much in relation to screen.	Set projector lens perpendicular to screen by adjusting projector or screen height and angle (see Overhead Projectors, InDepth I B, and Figure 3–2).
Picture Problems			
4. Poor illumination.	M, A	Room not dark enough.	Darken room further, or bring projector closer to screen.
		Projection lamp not properly inserted.	Check lamp placement.
		Old, dark-looking lamp.	Replace lamp.
		Wrong lamp.	Nameplate on projector may have lamp type. Otherwise, see Lamps.
		Greasy, dirty projection and condenser lenses.	Remove lenses. Wipe with lens tissue.

16mm Movie Projectors **147**

		STILL setting being used.	STILL setting cuts down lighting.
		Thin extension cord being used.	Use only heavy-duty cord.
5. Two half-pictures on screen.	M, A	Film not framed.	Turn FRAMER knob until there is only one picture on the screen.
6. Running blur; no image.		Film not properly set in aperture.	Open film gate. Check that film is properly set in aperture. Check that film is on sprockets. May need to rethread.
7. Cannot focus.	M, A	Lens barrel off gears so FOCUS knob, where available, will not work.	Push lens back into lens hole. Gears usually will engage. Use FOCUS knob or lens barrel to focus.

Figure 17-2 *When sound is garbled... (projector shown is Bell & Howell Autoload).*

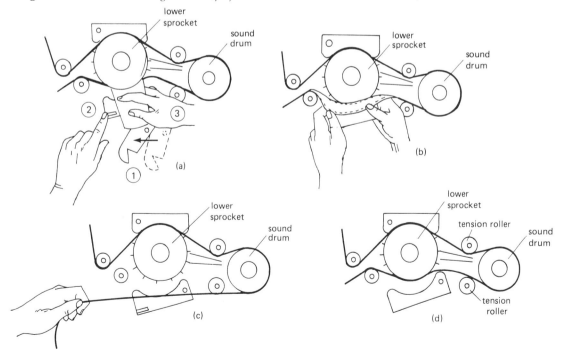

16mm Movie Projectors

PROBLEM	MANUAL (M) OR AUTOLOAD (A)	CAUSE	SOLUTION
Picture Problems			
		Cracked projection lens.	Replace projection lens.
8. Fuzzy picture.	M, A	Picture out of focus.	Use FOCUS knob.
		Cracked lens.	Replace lens.
		Greasy, dirty projection and condenser lenses.	Remove lenses. Wipe with lens tissue.
		Condensation on projection lens when projector is brought in from cold.	Wipe lens gently with clean facial tissue. Let projector warm up twenty to thirty minutes.
		Old, brittle film.	Cannot repair.
9. Momentary fuzzy picture.	M, A	Bad splice or torn sprocket holes.	Cut bad film, splice, and rethread projector. If film was borrowed, place flag in film reel at damaged portion. Include bad film when returning reel to owner.
10. Picture in and out of focus.	M, A	Film gate/aperture problem.	Take in for repair.
		Film not seated in aperture.	Rethread or open and close film gate.
		Film old.	Cannot repair.
11. Constant spots in one area of screen.	M, A	Greasy, dirty projection lens.	Remove lens. Wipe both ends of lens with lens tissue.
12. Fuzz projecting into picture.	M, A	Dirt in aperture.	Stop projector. Lightly clean aperture, using lens brush supplied with projector or cotton swab and alcohol.

13. One side of picture sharp; the other side out of focus.	M, A	Film not completely set in aperture.	Rethread film.
		Cracked projection lens.	Replace projection lens.
14. Film burning.		Heat filter missing.	Replace heat filter.
		Wrong lamp.	Nameplate on projector may have lamp type. Otherwise, see Lamps.

Sound Problems

15. Machine-gun-like popping sound when exciter lamp lights.	M, A	Silent film being used.	Check film. If sprocket holes on both sides, film is silent.
16. No sound; exciter lamp lights up.	M, A	If detachable speaker, cord or projector jack trouble.	Have service person check equipment.
		Film not threaded on sound drum correctly.	Using hand, push film back onto guides next to sound drum (see Figure 17–1, page 145).
		Dirt, dust, oil on sound drum.	See InDepth VI C.
		Motor switch not on FORWARD.	Switch motor to FORWARD.
		Sound switch not on.	Turn sound switch to ON.
		VOLUME low.	Raise VOLUME.
		Wrong type of projector for sound track; or magnetic sound switch not used on projector that has both optical and magnetic sound capabilities.	Use magnetic sound projector. See Glossary, magnetic sound track, optical sound system.
		Defective amplifier.	Take in for repair.

PROBLEM	MANUAL (M) OR AUTOLOAD (A)	CAUSE	SOLUTION
Sound Problems			
17. No sound; exciter lamp does not light up.	M, A	Exciter lamp burned out.	Replace lamp.
		Exciter lamp not properly inserted.	Check lamp placement.
		Wrong exciter lamp.	Nameplate on projector may have lamp type. Otherwise, see Lamps.
		Dual optical/ magnetic sound on magnetic sound setting.	Use optical sound switch (see Glossary).
		Sound switch not ON.	Turn sound switch to ON.
		Motor switch not on FORWARD.	Switch motor to FORWARD.
		Blown fuse.	Take in for repair.
18. Garbled, fuzzy sound, low volume (see Figure 17–2, page 147).	M, A	Too much slack in film; film's sound track cannot be decoded correctly by projector sound system.	Turn projector OFF. Make sure film is tight over sound drum. For manual load projectors, rethread film. For autoloads, make sure threading lever is in RUN position. (See Figure 17–2, a–1.) Release film from bottom sprocket and remove excess film, as follows: Adapt these directions to your unit: Open bottom fender under

16mm Movie Projectors **151**

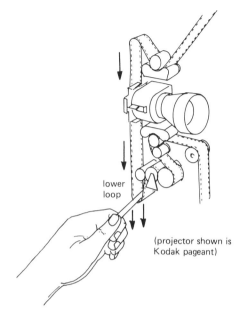

Figure 17-3 *When the speaker's lips on film and sound you hear don't match...(projector shown is Kodak Pageant).*

bottom sprocket gear. (Figure 17–2, a–2) With right hand, slide unused 3-inch by 3-inch index card under film (Figure 17–2, a–3).

Holding onto card and film, gently pull both outward as if removing film on a tray (Figure 17–2, b). Tighten tension rollers by pulling loosened film toward left (Figure 17–2, c). Reinsert taut film onto sprocket. Rollers should float. Close fenders (Figure 17–2, d).

	M, A	TONE not set properly.	Change TONE setting.
	A	Film not fully on sound drum.	Using index finger, push film back onto rollers and guides next to sound drum (see Figure 17–1, page 145).
19. Sound and picture not synchronized (lips do not correspond to sound. See Figure 17–3.)	M, A	Loss of lower loop.	While projector is running, momentarily pull down lower loop using pencil, or rethread film.

Film Problems

20. Film does not thread properly.	A		STOP PROJECTOR. Put motor switch on REVERSE, so film will exit at starting point. Determine cause, below. Reinsert film.

152 16mm Movie Projectors

PROBLEM	MANUAL (M) OR AUTOLOAD (A)	CAUSE	SOLUTION
Film Problems			
		Film not engaged by projector.	Manually push film into machine as far as it will go before automatic threading starts.
		Motor switch not on FORWARD.	Switch motor to FORWARD.
		Film leader not trimmed.	Use built-in trimmer.
		Film leader damaged.	Cut damaged portion; then use built-in trimmer.
		Lens housing not closed.	Close lens housing.
		Film path has film trimmings.	Check and clean film path.
		Projector not in THREADING position.	Set projector at THREADING position.
		Dirt in aperture.	Lightly clean aperture with lens brush supplied with projector, or use cotton swab and alcohol.
		Defective automatic film feeder.	As film feeds into machine, use your hand to press down curved plate to left of top sprocket. This applies particularly to Bell and Howell autoloads (see Figure 17–4).

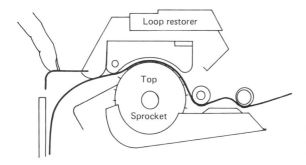

Figure 17-4 *One way to help machine feed film when automatic film feeder is defective.*

21. Projector makes chattering, clicking noise; film flickers.	M, A		STOP PROJECTOR IMMEDIATELY.
		Upper or lower loop too small.	Try each solution in turn: 1. Use loop restorer. 2. See Problem 19. 3. Allow a couple of extra frames through top sprocket (see InDepth IIIC and Figures 17–10a, b, c). 4. Rethread film.
		Lens housing not fully closed.	Close lens housing.
		Film gate coated with residue from new films.	Clean gate until shiny using alcohol and cotton swab.
		Sprocket holes chewed up.	Cut bad part of film, splice, rethread projector. If film borrowed, place flag in film reel at damaged portion. Include bad film and return with reel.

153

154 16mm Movie Projectors

PROBLEM	MANUAL (M) OR AUTOLOAD (A)	CAUSE	SOLUTION
Film Problems			
22. Film Scratched.	M, A	STOP PROJECTOR IMMEDIATELY.	Check projector and film, as explained below.
		Dirt or emulsion accumulated in film path or projector.	Clean film path (see InDepth VI).
		Film itself scratched.	No cure.
		Worn parts on projector.	Take in for repair.
23. Film breaks, jams; film piles up in projector.			STOP PROJECTOR IMMEDIATELY. Open sprockets and film gate. Remove film. For autoloads, see In-Depth IIIC.
		Poor splice.	Resplice film. Repair film with special splicing tape or special cement.
		Film reels not snapped into projector.	Repair film. Do not use tape, pins, or clips. Snap in reels.
		Sprocket holes chewed up.	Cut bad film, splice, rethread projector. If film borrowed, place flag in film reel at damaged portion. Include bad film and return with reel.

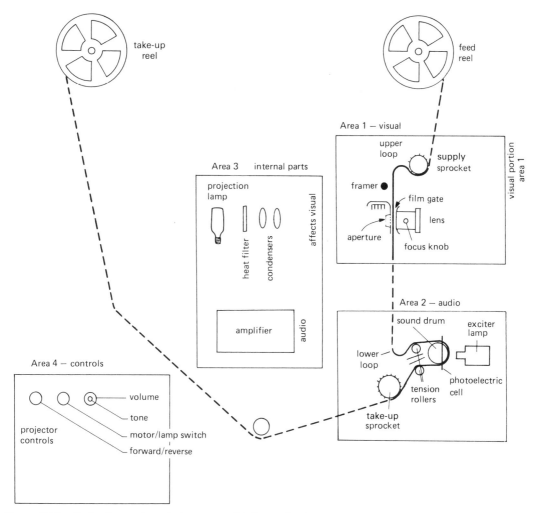

Figure 17-5 *Parts of a movie projector you need to know about.*

IN-DEPTH

I. Parts of a projector

Figure 17-5 labels the major parts of a 16mm projector and gives you an overview of threading procedures.

In general terms, the film follows this path:

from the supply or feed reel
to —> the supply sprocket
to —> upper loop
—> film gate/aperture (behind projection lens area)
—> lower loop
to —> the sound drum (past the projection lens)
to —> the take-up sprocket
to —> the take-up reel.

Projectors have guides, rollers and open-and-close sprocket clamps to guide and keep the film on its path.

155

Projectors often have threading diagrams on the projector cover. Some have a threading line for you to follow on the projector path itself.

Examples are given below for slot loading, automatic loading, and manual threading, which you can adapt to your units.

II. **Threading slot-load projectors**

  A. Recent models (examples: Eiki SL series, Bell and Howell 1575A, Singer, and Telex models). There is very little to do!

   1. Raise the projector reel arms and load the film reels.
   2. Switch the master control lever to LOAD, where applicable.
   3. With your hands, lay or guide the film in the film path and wind it around the take-up reel.
   4. Switch master control lever to PROJECT, where applicable.
   5. Select type of sound, where the projector provides a choice between optical or magnetic sound track.
   6. Activate the machine.

  B. Older models (example: older Graflex units. See Figure 17–6.)

   1. Raise the supply arms.
   2. Load the reels onto the projector (empty reel on the left and full reel on the right, with the film unwinding clockwise).
   3. Unwind five to six feet of film.
   4. Switch the master control lever to THREAD or LOAD (Push the REWIND control in, where applicable).
   5. Hold the first, or supply, sprocket open and insert film.

Figure 17-6 *Threading of slot-load projector.*

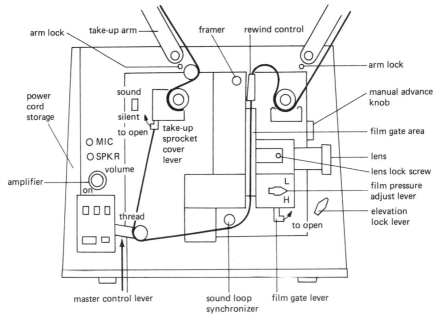

Make sure the sprocket holes are engaged in the sprockets.

6. Open the film gate. Move the film gate lever all the way to the right (located below LOW/HIGH lever).
7. Form a top loop approximately two fingers wide.
8. Insert the film into the film gate, making sure it is comfortably set. Images on the film should be upside down. Close the film gate.
9. Pass the film through the remainder of the film path, including the rollers. Do not pass the film on the master control lever itself. Carefully follow the projector markings for the lower loop position.
10. Open the take-up sprocket lever (located below the take-up arm). Wind the film around the sprocket, making sure the sprocket holes are engaged. Close the lever.
11. Bring the film to the take-up arm, winding it clockwise under the empty reel.
12. Where available, turn the manual advance knob (next to the lens) to check for correct threading.
13. Raise the master control lever to FORWARD or PROJECT. You are now ready to run the machine.

C. Slot-load threading precautions (mainly for older models).
1. The film should be tight around the sound drum; otherwise, the sound will be garbled.
2. The lower loop must be set properly for synchronization of picture and sound (see Problem 19).

III. **Threading autoload projectors (example: Bell and Howell. See Figure 17–7)**

A. Automatic threading.
1. This is crucial. Trim the film before inserting it into the projector. Use the built-in film cutter provided on either the machine or the cover (see Figures 17–7a and 17–8).
2. Raise the supply arms.
3. Load the reels onto the projector. Place the empty reel on the left and the full reel on the right, with the film unwinding clockwise (see Figure 17–7b).
4. Move the threading lever to the right (counterclockwise) until the mechanism clicks or otherwise prepare for automatic threading (see Figure 17–7c).
5. Turn the motor switch to FORWARD (see Figure 17–7d).
6. Insert the film into the machine (under roller "4" for Bell and Howell) until it is engaged (see Figure 17–7e).
7. When the film emerges where you can hold it, gently pull the film fully toward the left until the threading lever clicks open. Press STOP, on some projectors (see Figure 17–7f).
8. Finish winding the film onto the take-up reel (see Figure 17–7g).

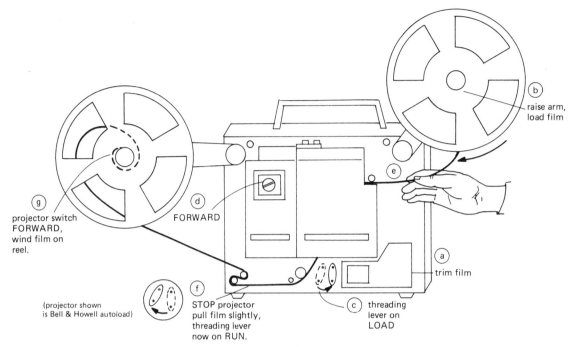

Figure 17-7 *Steps in threading Autoload (projector used is Bell & Howell Autoload).*

B. Manual threading (when starting a film in mid-reel).

1. Raise the supply arms for RUN position.

2. Load the reels on the projector. Place the empty reel on the left, the full reel on the right,

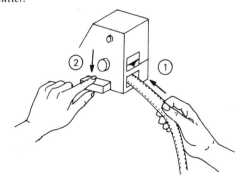

Figure 17-8 *Remember to trim film! Use projector cutter.*

with the film unwinding clockwise. Unwind a couple of feet of film from full reel.

3. Open the hinged lens housing, usually by pulling the lens housing toward you.

4. Move the threading lever to the left (clockwise). (For 5 to 8 below, see Figure 17–9.)

5. Open all sprocket guards or clamps. In example shown, push up on A and B, down on C.

6. Thread film. Follow the film path in the diagram. Allow the film to loop at X and Y until the film reaches the curved metal retainers. Check that the sprocket holes of the film are engaged in the sprocket wheel teeth.

158

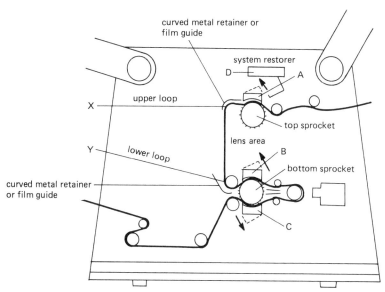

Figure 17-9 *Manual threading of Autoload projector (projector used is Bell and Howell Autoload).*

7. Close each of the sprocket guards—A, B, and C. Close lens housing. Wind film onto empty reel.

8. Test the projector. Turn the motor switch to FORWARD. If picture and sound are not synchronized or the film flickers, the loops may be the wrong size. Correct as follows:

   - On the unit in the diagram, press down on the System Restorer (D on the diagram) while the switch is on FORWARD, or

   - Open the sprocket guard at A and remove the film from the sprocket. Allow film slack for the larger top loop. Replace the film in the sprocket and close the sprocket guard. For bottom loop, see Problem 19.

C. Unthreading film before it is finished

If the film gets stuck in the machine and has to be removed, first try switching the motor/lamp to REVERSE. Otherwise, remove manually. Adapt the directions to your machine. You may want to pay particular attention to directions and diagrams for removing film from the sprocket wheels. (See Figure 17–10.)

1. Open the lens housing. In some cases, also remove exciter lamp cover.

2. Turn the supply reel clockwise to provide slack in the film. Also make sure the threading lever is in the RUN position. See Figure 17–7c.

3. Remove film from first or supply sprocket wheel. This sprocket is usually located at

159

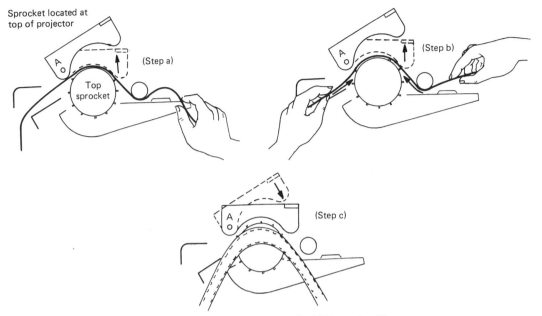

Figure 17-10 *Unthreading Autoload before film ends. (a), (b), (c) Removing film from top sprocket. (d), (e), (f) Removing film from bottom sprocket.*

the upper part of your projector.

Use your right hand to ease out the film from under roller and hold onto the film. (See Figure 17–10a.)

With your left hand, reach in and hold the film on the left side of the sprocket wheel. (See Figure 17–10b.)

Push your hands *toward each other* and ease the film off the sprocket teeth. (See Figure 17–10c.)

4. Remove the film from the *top* of the second or take-up sprocket wheel. This sprocket is often located either at lower area or at left side of projector.

Push up at top fender.

Put your right index finger on top of the film under the roller next to the sound drum. Hold the film at the lower loop with your left hand. (See Figure 17–10d.) With both hands, push upward or slide the film *toward the sprocket* wheel. In the same motion, gently remove the film from the top sprocket teeth and the top roller. (See Figure 17–10e.)

5. Remove the film from the bottom of the second or take-up sprocket wheel.

Open bottom fender. Use your right hand to hold the film below the sound drum. Hold the film left of the sprocket with your left hand. Gently push upward and in the same motion remove the film from the bottom teeth of the sprocket. (See Figure

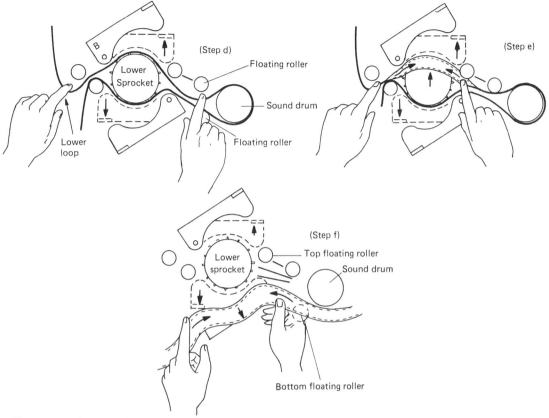

Figure 17-10 (*continued*)

17–10f.) Slide the film out of the bottom film channel. Rewind.

6. Replace the exciter lamp cover if removed; close the lens housing.

IV. Threading manual projectors (See Figure 17–11)

A. Raise the supply arms.

B. Load the reels on the projector, the empty reel on the left, the full reel on the right, with the film unwinding clockwise.

C. Unwind 5 to 6 feet of film.

D. Open all guards or clamps above and below the sprockets; open the film gate where necessary for threading.

E. Follow the threading pattern. (See Figure 17 5.) Here are some general threading guidelines:

Form upper and lower loops. Film should be well seated in aperture area.

Make sure the film is on the sprocket teeth.

Close the sprocket guards and the film gate as you go.

F. Do not allow film slack on sound drum. Also, film should be pushed fully onto sound drum.

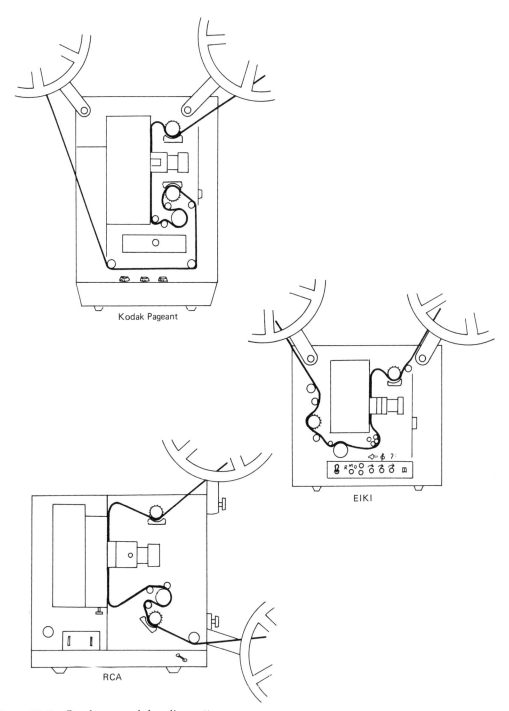

Figure 17-11 *Popular manual threading patterns.*

G. Wind the film onto the take-up arm, winding it under the empty reel clockwise.

V. Rewinding film

A. On most projectors, the film should be out of the projector film path. Manually remove the film in autoload projector if the reel is not finished (See InDepth III C.)

B. For some projectors, when film is out of path raise the arm of the left reel to the upright position.

C. Place the full reel on the left, the original empty reel on the right.

D. Insert film from full reel directly onto the original reel so it winds counterclockwise.

E. Activate the rewind mechanism appropriate for your unit.

F. Turn the motor switch to OFF.

VI. Cleaning your projector

Three areas need cleaning:

A. *Film path.* Clean the film path with the brush that is provided with the projector, or a toothbrush. Then use a soft, clean cloth dampened with alcohol. Clean the aperture area and the film gate regularly with cotton swabs dipped in alcohol until both are shiny.

B. *Lens.* The projector should be cooled off before cleaning the lenses. The projection lens and accessible condenser lenses should be cleaned regularly on both ends with lens tissue or a soft, dry cloth. Do not use facial tissue.

C. *Sound area.* For optical sound system projectors, there is an exciter lamp, a sound drum, and an amplifier and speaker. Do not touch the speaker. The exciter lamp and the sound drum need cleaning. In some cases, removing the exciter lamp cover provides more complete access to the sound drum. Use a soft brush, or use a new pipe cleaner to clean under the edge of the sound drum or other hard-to-reach areas.

For projectors with magnetic stripe sound system, the sound head is similar to a tape recorder head. Clean with cotton swab and alcohol or cassette head cleaning fluid.

VII. Running times for various size reels

Table 17–1

Size of Reel (diameter in inches)	Full Reel of Film (in feet)	Running Time (minutes)
7	400	11.7
10½	800	22.13
12	1200	33.20
14	1600	44.30

Note: 50 feet of film = approximately 1½ minutes of running time.

GLOSSARY

Aperture Opening in the projector that permits light from the lamp to strike the film.

Condenser lenses Lenses that concentrate light through the image to be projected. Located inside the projector.

Emulsion Dull-looking side of film. Contains the image. Can easily be damaged. Do not touch.

Exciter lamp Also called sound lamp. See Optical sound system.

Feed reel Full reel of film.

Film channel See Film path.

Film cutter or trimmer Device usually found on the side or base of the projector or on the projector cover. Used to clip film before threading the projector. Particularly important for autoload machines.

Film gate Track and part of the lens housing that holds film against the aperture in front of the projection lamp while the film is projected.

Film path Entire path the film follows from feed reel to take-up reel. Also called *film channel*.

Flag Bookmark to identify a particular spot on a reel of film. It can be a small scrap of paper. Do not use clips.

Framer Knob used to obtain one complete picture on the screen instead of halves of two pictures.

Leader About 5 to 6 feet at the film's beginning or end used to ease threading or rewinding. It may be clear-looking or colored and does not have film content.

Lens Projection lens. Permits focus adjustment of the picture on the screen.

Lens housing Hinged portion of the projector in which the projection lens is mounted. Can be opened to check film in the film gate, to clean the aperture area, and to clean the back of the projection lens.

Loops Extra lengths of film at designated places in the film path. Without proper loops, the picture will flicker and sound and picture will not synchronize.

Loop restorer Device used when the film is not feeding properly through the projector. Used while the projector is in FORWARD.

Magnetic sound system Sound system for use with sound track that has a stripe of recorded magnetic tape running down the film edge. Used on super 8 sound cartridges and reel-to-reel films. Not as commonly used with 16mm films. The stripe looks brown and looks like and operates like audio tape.

Optical sound system Sound system for use with sound track that has been photographed on film. The fine wiggly line you see when holding a piece of film is turned back into sound as the film passes through the projector. In general terms, the projector exciter lamp beam shines on the film sound track and onto a photoelectric cell. The cell converts the light into electrical impulses. The projector amplifier strengthens these impulses, which are turned into sound through the speaker. This sound system is in common use with 16mm projectors. Some super 8 and 16mm projectors have a dual optical sound/magnetic sound switch to adapt to the particular film you are using.

Projection lamp The large lamp used to project the image on the screen.

Pulldown claw Pulls each frame of film past the aperture as it is being projected.

Reverse Projector function that makes the

film operate backward. Should only be used for a short stretch of film.

Rewind Projector function that puts the film back onto the original reel the film came with after projection has been completed. Usually takes place with film out of the film path.

Sound drum A wheel over which film passes at a set rate of speed.

Sprocket guard Usually, plastic lever that protects the film and guides it onto the sprocket wheel.

Take-up reel Empty reel.

Take-up sprocket Transports the film out of the projector onto the take-up reel.

Threading lamp Found on some projectors, a small lamp used for threading or checking the projector after the house lights have been turned off. When the lamp is burned out, ask for a new one by the model number of the projector.

18 Tape Recorders

CONTENTS

Operating Tips
- General 167
- Tape 167
- Microphone 168

Problems
- Power Problems 169
- Playback Sound 169
- Recording 172
- Tape 172
- Synchronization 174

InDepth
Cassette Units
- Playback Procedures 175
- Recording Procedures 175
- Setting Recording Levels 176
- Some Special Features on Some Recorders 176
- Erasing Procedures 177
- Erase Protection Device 177
- Batteries 177
- Using Playback Unit in AC Mode 177
- Cleaning and Demagnetizing Recorder Heads 178
- Cassette Tapes 178
- Editing Tapes 179
- Synchronizing Cassette Recorder with Slides, Filmstrips, or Other Machines 179
- Duplicating Tapes 180

Reel-to-Reel Units
- Playback Procedures 180
- Recording Procedures 182
- Setting Machine Speed 182
- Activating Recording Mechanism 182
- Setting Recording Levels 183
- Tapes 183
- See *Cassette Units* for:
- Some Special Features 183

Glossary 184

CASSETTE UNITS

Cassette units record and play back sound. There has been an increased use of cassette recorders in the last few years because of their portability and ease of handling. They are usually standardized in speed and tape format. While cassette recorders include the playback mechanism, cassette players are listening devices and cannot record. Cassette recorders and players vary greatly in model, size, fidelity, style, and price.

166 *Tape Recorders*

The problem with this type of equipment is making major repairs on less expensive machines. Parts for older machines are difficult to find, and often it is not worth the cost to repair an inexpensive recorder.

REEL-TO-REEL UNITS

Reel-to-reel units record and play back sound. They have controls and jacks that are similar to those on cassette units. The machines tend to be larger. Reel-to-reel recorders usually do not have battery operation and, unlike most constant-speed cassettes, they can be operated at one of several speeds, according to the use intended. The time a given reel records or plays depends on the speed used.

OPERATING TIPS

I. General

YOUR TAPED SOUND CANNOT BE BETTER THAN THE ORIGINAL SOUND.

A. There are six standard controls used to operate a tape machine: PLAY, FAST FORWARD, RECORD (for recorders), REWIND, STOP, and EJECT. Some machines also have a PAUSE control.

B. When changing the function (as from REWIND to STOP), usually press the buttons or turn the switches firmly. Where applicable, make them click and not slide gently.

C. Some machines have VU meters or LEDs to register sound level. You are recording if the tape is turning and the meter or LED, where available, is working.

D. On most machines, press STOP before changing functions. On some machines, press STOP even when the tape ends and stops; otherwise, the machine motor keeps running.

E. Always test the recording level before recording. Levels differ in location and individual voice. (See In-Depth III for Cassette Units, and In-Depth V for Reel-to-Reel).

F. Periodic head cleaning on machines is essential to avoid sound problems (see InDepth IX, Cassette Units).

G. Keep the machines away from heat or moisture.

H. Buy alkaline batteries, where batteries are needed.

I. When inserting the AC cord into the machine, note that it should only go in one way; otherwise, the machine will be damaged.

J. The automatic level control (ALC or ARL) is for recording; it does not work for playback levels.

K. Some cassette recorder-players have an AC/battery switch. If you are using a machine with one type of power, press STOP before changing the power source.

L. Remove the batteries from the cassette recorder when they are weak or when using AC power for long periods.

M. For reel-to-reel units, the listening or playback speed must be the same as the recording speed.

N. Some cassette units have a variable listening speed. This increases or decreases the playback speed of a tape recorded at normal speed.

II. Tape tips

A. The machine capability, not the kind of tape, determines the number of tracks and the direction in which they can be recorded and played. Recording or playback can only occur on one side of the specially coated tape.

B. Do not store tapes in direct sunlight or near heat sources. Heat dries out the tape, which then becomes brittle or warped. Store tapes in a cool place.

C. There is no special way to store old, brittle tape. However, a recording or tape-duplicating professional may be able to transfer it to a new tape.

D. Do not put tapes near moisture.

E. Do not place a recorded tape near a microwave oven, a radio, a TV, or any other magnetic force, as this may affect tape sensitivity and may erase the recording.

F. A new tape may be soft or sticky. Run the entire length of the tape through the machine on FAST FORWARD at least once, and tap the tape on a table before inserting it in the machine.

G. Before inserting a tape into the machine, take up any slack in the tape by turning a pencil in the hub of the reel. This will prevent possible tape breakage when playing the tape.

H. Do not use 90- or 120-minute cassette tapes if possible. They break more frequently than 30- or 60-minute tapes because the tape is thinner.

I. Do not use cheap or off-brand tapes. They may not be well made, and may therefore cause machine problems and poor sound reproduction. Here is one way to tell if they are poorly made: Lightly rub the dull side of the tape with a pencil eraser. If residue comes off easily on the eraser, the tape is cheap!

III. **Microphone tips for remote microphone**

A. Make sure the remote microphone and the tape recorder have matching impedance (see Microphones).

B. Some remote microphones have a mike switch and a cord with two plugs to connect the microphone to the recorder. When plugged into the recorder the switch and the smaller plug control the recorder's STOP/START (ON/OFF) mechanism. The larger plug controls the microphone input signal. You may use this kind of microphone even if the recorder has only one microphone jack instead of two. Just connect the larger plug to the recorder's MIC jack to record the sound.

C. When using automatic level control (ALC), you cannot adjust the recording volume. Bringing the source of sound closer to the microphone will diminish background noise.

D. Keep the microphone as far away as possible from the recorder.

E. Hold the microphone three inches slightly to the side of, and six to ten inches away from, your mouth. Otherwise, "b" and "p" sounds will pop.

F. Where available, turn MONITOR or SPEAKER switch OFF while recording, unless you are using headsets.

For more information, see Microphones chapter.

PROBLEMS

For best results, test equipment before using. *FIFTY TO NINETY PERCENT OF ALL TAPE FAILURES ARE CAUSED BY DIRTY HEADS!* Sound quality will be poor if the recording equipment and/or playback equipment has dirty heads. To clean dirty heads, see InDepth IX, Cassette Units.

PROBLEM	CAUSE	SOLUTION
Power Problems		
1. No power when using batteries.	Dead batteries. Can cause equipment damage.	Replace batteries or use AC.
	AC/battery switch set on wrong power source.	Flip switch to proper source.
	Detachable AC cord in machine.	Remove detachable cord from machine and AC outlet.
	Battery and/or machine contacts dirty.	Wipe battery contacts with rough cloth, and machine contacts with typewriter eraser.
	Batteries inserted incorrectly. Can cause equipment damage.	Check diagram in battery compartment.
2. No power, other causes.	Recorder/player switched OFF.	Turn machine ON where switch is available.
	Remote microphone plugged in but switched OFF.	Switch remote microphone to ON where switch is available.
	AC cord not plugged into machine or outlet.	Check AC connections.
	Cord defective, frayed, or improperly connected.	Check cord connections. Replace if necessary.
	Defective machine.	Take in for repair.
	For reel-to-reel, also check: Machine not threaded correctly.	On some machines, check that tape has threaded within automatic cut-off spool.
Playback Sound Problems		
3. No sound on playback (see also Problems 1, 2, 14).	Power OFF.	Turn power ON.

170 *Tape Recorders*

PROBLEM	CAUSE	SOLUTION
Playback Sound Problems		
	Volume level too low.	Raise volume level.
	PAUSE switch ON.	Turn switch OFF.
	Full reel on right side.	Change full reel to left spindle or rewind tape. Flip cassette tape to side 2.
	Defective machine.	Take in for repair.
	Defective tape.	Test on another machine. May be unuseable.

If none of the above applies, check problems during recording: power OFF, defective tape, microphone not suited for recorder, condenser mike battery missing or old, recording level too low, remote microphone OFF, PAUSE switch ON, defective machine. For reel-to-reel units, also check that the dull side of the tape is facing the heads. Locate and correct the problem. Record again.

PROBLEM	CAUSE	SOLUTION
4. Weak sound on playback.	Volume level too low.	Raise volume level.
	Too many headsets.	Use fewer headsets.
	Dirty heads.	Clean heads.
	Defective machine.	Take in for repair.
	For cassette units, also check: Weak batteries.	Test and replace.
	For reel-to-reel units, also check: Poor splice, weak sound at splice.	Redo splice; use only audio splicing tape.

If none of the above applies, check problems during recording: poor sound input, weak batteries, dirty heads, microphone not suited for recorder, recording level too low, remote microphone too far from sound input, defective machine. Locate and correct the problem. Record again.

PROBLEM	CAUSE	SOLUTION
5. Wow and flutter, distorted sound, noise on playback.	Dirty heads, capstan, rubber roller.	Clean where necessary.
	Defective tape.	Test on another machine. May be unusable.
	Defective machine.	Take in for repair.
	For reel-to-reel units, also check: Listening speed not the same as recording speed;	Check recording speed and adjust listening speed accordingly.

	Back side of tape or shiny side next to heads.	Tape now has half twist. Play or rewind tape adding your own half twist. Tape will correct itself.
	For cassette units, also check: Weak batteries;	Test and replace.
	Wrong AC adaptor on a playback-only unit that is battery-operated.	Get appropriate AC adaptor (see InDepth VIII, Cassette Units).

If none of the above applies, check problems during recording: poor sound input, dirty heads and rollers, defective tape, microphone too close to speaker's mouth, defective microphone, weak batteries, recording level too high. Locate and correct the problem. Record again.

6. Sound mushy. No high-frequency sound.	*Reel-to-reel units:* Back (shiny) side of tape being played next to machine heads.	See Problem 5.

Recording with the back side of the tape next to heads may have occurred. You may have to record again.

7. Fading in and out on playback.	Defective cord or connections.	Check connections.
	Dirty heads.	Clean heads.
	Tape partly erased.	Rerecord tape. Store tape properly (see Operating Tips II).

If none of the above applies, check problems during recording: poor sound input, defective cord or connections, dirty heads, defective microphone, defective tape, constant distance not maintained between sound input and microphone. Locate and correct the problem. Record again.

8. Humming or scratchy noise on playback.	Dirty heads.	Clean heads.
	Magnetized heads.	Demagnetize heads.
	Defective tape.	Test on another machine. May be unuseable.
	Defective machine.	Take in for repair.
9. Unwanted radio station programming on playback.	Interference from radio, TV, appliances.	Locate interference source; turn source off or relocate recorder away from source.

PROBLEM	CAUSE	SOLUTION
Playback Sound Problems		
	Machine not grounded properly.	If using AC without three-prong plug, reverse position of plug in AC outlet.

If none of the above applies, check problems during recording: poor sound input, microphone not securely connected, defective microphone, dirty heads, defective tape, interference from radio, TV, and other sources. Locate and correct the problem. Record again.

PROBLEM	CAUSE	SOLUTION
10. Previously recorded material did not erase completely when rerecording.	Dirty heads.	Clean heads.
	Defective machine.	Take in for repair.
11. Reel-to-reel recorders: Two different sounds, one normal, one backward.	Stereo tape being played on monaural machine.	Obtain stereo tape recorder.
Recording Problems		
12. RECORD button does not depress.	Recorder on PLAY.	Do not force record button. Press STOP.
		On most machines press RECORD and PLAY *together*; or, for some reel-to-reel units, locate record lock, release, and press RECORD without pressing PLAY.
	For cassette units, also check: No cassette in recorder;	Insert cassette. Press RECORD and PLAY.
	Break-out tabs on tape removed to prevent accidental erasure.	Cover holes with masking or cellophane tape. Press RECORD and PLAY.
13. No RECORD button on cassette unit.	Machine does not record; this is a playback-only unit.	Obtain recorder-player.
Tape Problems		
14. Tape stuck; will not move. (See also Problems 1,2.)	Tape caught above or below pinch or rubber roller.	Use tweezers to remove tape.

	Full reel on right side.	Change full reel to left spindle or rewind tape. Flip cassette to side 2.
	Defective machine.	Take in for repair.
	Defective tape.	Replace tape.
	Mode not selected.	Press PLAY for playback, or RECORD and PLAY for recording.
	For reel-to-reel units, also check: Tape not properly threaded.	Check threading (see Figure 18–3).
15. Tape squeal or squeak.	Dirty heads on playback or during recording.	If playback problem, clean heads and play tape. If recording problem, clean heads and record again.
	Poor-quality tape.	Change brand of tape. Record again.
16. Cassette tape runs backwards on AC power.	Wrong AC adaptor for cassette unit.	Get proper adaptor (see InDepth VIII).
17. Tape unthreads and spills out of housing while machine is running.	Defective tape, or dirty capstan pressure roller.	Stop machine. Rewind tape with a pencil in hub. Clean capstan roller with alcohol and cotton swab. Duplicate tape. Keep original, but use rarely.
	Defective machine.	Rewind tape with pencil in hub. Take machine in for repair.
	Take-up reel not moving.	Remove foreign object touching reel.
		Belts worn. Take in for repair.
	Used longer length cassette tape, 90 or 120 minutes.	Try to duplicate tape on 30- or 60-minute tapes.

174 *Tape Recorders*

PROBLEM	CAUSE	SOLUTION
Synchronization Problems		
18. Tape not automatically synchronizing or controlling other machines as planned.	Patchcords connecting all equipment not plugged in.	Check all patchcord connections.
	Tape recorder not designed to control other machines.	Use special equipment (see InDepth XII A, Cassette Units).
	Tape not programmed with special equipment.	Operate other machines manually to coordinate with taped comments (see InDepth XII B, Cassette Units).
	Taped signal does not work on your tape recorder.	Check whether tape and tape recorder were meant for use with slides. See Slide Projectors, InDepth VIII D. Obtain proper machine, or manually operate projectors with taped comments. See Slide Projectors, InDepth VIII C.
	Defective machine.	Take in for repair.
	Defective patchcord.	Replace patchcord.
	Tape and visual not started simultaneously.	Rewind tape and visual. Start both machines at same time at appropriate spot.
	Tape advanced or rewound but visual not made to match.	Advance or reverse visual to match tape.
	Visual advanced or reversed but tape not made to match.	Advance or rewind tape to match visual.

See also Slide Projectors, Problems 18, 19; Filmstrip Projectors, Problem 18; and Dissolve Unit, Problems 21, 22.

IN-DEPTH—CASSETTE UNITS

(See Figure 18–1.)

I. **Playback procedures**

A. Set up the machine for AC or battery operation. Plug in the AC cord or check the batteries.

B. Turn the power ON, where such a switch is provided.

C. Set the index counter, where available, to "000."

D. Press the EJECT button. Insert the cassette with the side to be listened to facing up and the full reel of tape on the left. Close the cassette compartment.

E. Press PLAY. Set volume level.

F. Press STOP when finished. Use FAST FORWARD and/or REWIND to locate material on the tape, using the index counter as a guide.

G. If not using AUTOREVERSE (see Glossary), when finished rewind the tape to the beginning of side 1 for playback, or flip the cassette and listen to side 2.

II. **Recording procedures**

A. Set up the machine for AC or battery operation. Plug in the AC cord or check the batteries.

B. Turn the power ON, if such a switch is provided.

C. Press the EJECT button. Insert the cassette with the side to be recorded facing up and the full reel of tape on the left. To check that the reel is completely rewound, press REWIND.

D. Set the index counter to "000" where available.

E. Plug the microphone into the recorder's MIC jack if there is not microphone built in or if you want to use a separate microphone (see also Operating Tips III).

F. On many recorders, press RECORD and PLAY simultaneously.

G. Turn the microphone switch ON, if

Figure 18-1 *Parts of a cassette recorder/player unit*

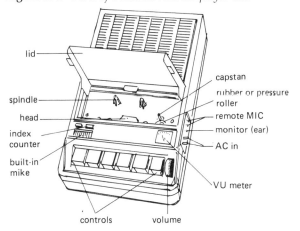
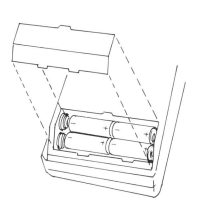

a switch is provided on a microphone not built into the recorder.

H. Set the recording level, if your machine provides that option (see InDepth III).

I. Stop the tape. Rewind the tape, then press PLAY for playback of the tape to check the recording level and to test the equipment.

J. If the sound level is satisfactory, rewind the tape, stop, and then, on many machines, press RECORD and PLAY simultaneously to record the program.

K. Allow the tape to run for about ten seconds, making sure the leader tape has disappeared onto the right reel. Then begin recording.

L. When recording on side 1 is finished, press STOP.

M. Rewind the tape for playback, or turn the cassette to side 2 and continue recording. If you did not use all the tape on side 1, press REWIND on side 2 before recording.

III. Setting recording levels

Today most recorders automatically and electronically adjust the volume level while recording, according to the loudest sound. Therefore, in many cases the VOLUME knob is to adjust playback, not recording volume.

A. Automatic setting (called ALC or ARL on some recorders).
Set the switch on the recorder to ALC or ARL, where available. The recorder will adjust the recording level automatically, according to the loudest sound.

B. Manual setting.
Set the switch to MANUAL or turn ALC to OFF, where the recorder has both capabilities. Press RECORD and PLAY.
Speak into the built-in recorder microphone or a separate microphone. Adjust both the recording level knob and as you talk your distance from the microphone until the needle does not peak into the red area on the record level meter, or the LED level does not peak on the scale. Where there is no meter on the recorder, usually assume it is ALC only.

IV. Some special features on some recorders

A. Monitor switch.
This switch allows the user to hear what is being recorded while the machine is on RECORD/PLAY. Set the switch on the recorder to MONITOR, or turn the monitor switch ON, whichever is applicable. Plug an earphone or a headset into input by that name. Turn switch OFF when not using earphones. The monitor switch ordinarily does not function on playback.

B. ON/OFF switch on remote microphone, or PAUSE (STANDBY) on recorder.
This is useful when the tape must be stopped and restarted many times. The ON/OFF microphone switch should be used primarily during recording. On some units, you must unplug the microphone during playback. The PAUSE switch can be used during either recording or playback.

V. Erasing procedures

There are three ways to erase a tape:

1. Insert the cassette. Record new material, using normal recording procedures.
2. Insert the cassette. Press RECORD and PLAY. Do not plug in the remote microphone. If there is a built-in microphone, buy a dummy plug and insert it into the MIC jack.
3. Use a bulk eraser, a large magnet over which the cassette is slowly turned. Buy a bulk eraser from a cassette dealer.

VI. Erase protection device

All cassette tapes have erasure protection tabs to prevent accidental erasure (see Figure 18–2). To make the recording permanent, punch out the tabs by applying slight pressure with a key. The record button cannot be depressed with the tape tabs punched out. To rerecord on a cassette with broken tabs, cover the holes with masking or cellophane tape and record as usual. Holding cassette with side 1 facing up, the left tab affects side 1, and the right tab affects side 2 or the bottom side.

VII. Batteries

Checking batteries:

You may check the batteries in the recorder through a built-in meter, where available. Depending on your unit, press RECORD or PLAY, to get a battery reading.

If the batteries are weak, you may

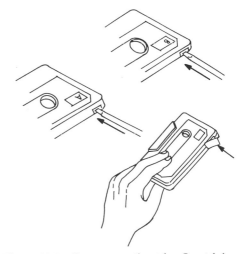

Figure 18-2 *Erase prevention tabs. One tab for each side of tape. For side 1 or A punch out left side as tape faces up. Repeat for side B. To record again, cover hole with tape.*

use them for playback but not for recording.

A drugstore, photo dealer, or hardware store can check the batteries if you are in doubt.

For battery care and cleaning, see chapter on General Tips, Batteries.

VIII. Using playback unit in AC mode

While recorder-players often come equipped for use with either AC or batteries, some playback-only units do not come with a built-in AC cord and plug; a separate AC adaptor is therefore necessary. An AC adaptor is a battery eliminator. Buy the appropriate one for your make and model machine. A universal adaptor can be used with most units. Adaptors are relatively inexpensive and are available at cassette dealers. Remember to unplug the adaptor when the recorder is not in use on AC.

IX. **Cleaning and demagnetizing recorder heads**

A. General comments.

The heads of the recorder decode the sound on the tape: They become dirty with dust and oxides deposited by the tapes. Even a 0.0001-inch buildup will affect performance. One defective tape can get heads or rollers dirty. Heads, therefore, need to be cleaned and demagnetized to remove the oxides.

Clean heads and rollers for the first time after twenty hours of use and after every forty hours of use thereafter.

If playing or recording a valuable tape, clean the recorder heads before using the machine. You will not damage the heads by cleaning them often if cleaning is done properly.

Demagnetize after every 100 hours of use. The quality of the tape will be degraded if the machine is magnetized. Make sure the power is turned OFF when demagnetizing the recorder.

B. Cleaning heads and roller by hand.

1. Remove the cassette tape.
2. Press the PLAY button to push out the heads so they are accessible to you. On a few models, PLAY will not depress completely. Do not force.
3. Never touch the heads with a sharp object, a stiff brush, or anything metallic.
4. Wipe the heads with a cotton swab moistened, not soaked, with alcohol. Do not wrap the swab in anything metallic, as this might scratch or magnetize the heads.
5. Use the same solution for the capstan roller and the rubber roller.

C. Using head-cleaning tape (for cassette recorders).

Use this tape especially for machines where the heads are difficult to reach by hand.

1. Purchase head-cleaning cassette tape.
2. Insert into the recorder. Set the index counter at "000."
3. Press PLAY.
4. Let tape run for about one minute, or follow package directions. Press STOP.
5. Remove the tape.

This tape only cleans heads. Clean the capstan and rubber rollers as in InDepth IXB above.

D. Demagnetizing heads.

Use a head demagnetizer available at your dealer. Instructions accompany the device. However, it can damage the equipment if it is not used carefully.

X. **Cassette tapes**

A. Cassette tapes or cassettes, as they are commonly called, consist of two reels of tape in a housing. The entire cassette is placed in the recorder. All housings are standard in size, regardless of the length of the tape.

B. When purchasing cassettes, look for the following:
1. *Quality.* Cassettes vary in price according to the quality of the tape, format, and brand name. Purchase only well-known brands to be assured of good quality. Cheap tape can cause machine problems and poor sound reproduction.
2. *Appropriate tapes.* Use tape suitable for your machine. For example, metal tapes should be used only with equipment built to handle them.
3. *Cassette housing.* Cassettes can be bought permanently sealed or with five screws. Those with screws are more expensive.
4. *Bulk purchasing.* Buying in bulk quantities is very economical and, in some instances, can save you 50 percent or more.

C. Formats

The most common cassette formats are:

- C-15   7½ minutes on each side = 15 minutes
- C-30   15 minutes on each side = 30 minutes
- C-60   30 minutes on each side = 60 minutes
- C-90   45 minutes on each side = 90 minutes
- C-120 60 minutes on each side = 120 minutes

The 90- and 120-minute cassettes are not recommended for general use or for high-speed duplicating because the tape is too thin and often stretches, jams or breaks.

Cassettes are also available in 2-minute increments: 2 minutes, 4 minutes, 6 minutes, and so on. However, the cost is great. For example, it may only cost 15 percent more to buy a 10-minute tape than a 2-minute tape. Figure out the most economical price for your intended use.

Tapes in a continuous loop never need to stop. These are more expensive than standard format cassettes.

Better-quality cassette tapes are usually available in low-noise formats, which are recommended for use in recording music, and other situations where high fidelity is necessary.

XI. **Editing tapes**

Editing cassette tapes is difficult because the tape is so narrow. It can be done, though, and special cassettes and splicing tape are available. However, editing and splicing cassette tape is not recommended. Instead, edit a program on reel-to-reel tapes, then transfer the sound back to a cassette format.

XII. **Synchronizing cassette recorder with slides, filmstrips, or other machines** (see also Slide Projectors, InDepth VIII, and Filmstrip Projectors, InDepth III).

A. Automatic advance.

Some more expensive recorder-players have a special sync feature. During the recording process, the tape is specially programmed with inaudible electronic signals. On playback, when the tape recorder is connected to the appropriate machines it is designed to control, the tape automatically advances or stops such machines. It can control slide projectors, filmstrip projectors, or other machines as well as the tape recorder itself. When a cassette machine is connected to other equipment, a special patchcord must be used. It can be obtained from an audio/visual dealer.

B. Manual advance.

For a recorder that does not have the special sync feature to record an electronic signal on the tape, you may record an audible signal as you record your script. Simply clap your hands, touch a spoon to a glass, or use a dime store clicker at appropriate intervals. On playback, this audible signal will tell you when to advance visuals or perform other functions called for in the script. If there are no signals, follow the script during your presentation.

XIII. Duplicating tapes

A. Using the high-speed duplicator. This machine can handle one master and make several copies at the same time. It takes only a few minutes, depending on the length of the original. Some duplicators can transfer format—from a cassette to a reel or from a reel to a cassette.

The high-speed duplicator can handle any good-quality tape. Do not duplicate tapes that have been spliced or damaged, or cassette tapes that are 90 or 120 minutes.

B. Using two cassette or reel-to-reel machines—one with original, the other with copy.

Use any good-quality brand-name tape. You may duplicate tapes not acceptable in a high-speed duplicator. This process takes as long as the time on the original tape. A 30-minute tape would take 30 minutes of duplicating time.

See also chapter on Connectors, Jacks, and Plugs.

IN-DEPTH—REEL-TO-REEL UNITS

(See Figures 18-3 and 18-4.)

I. Playback procedures

A. Plug the cord into an AC outlet.

B. Remove the spindle protectors, if there are any.

C. Place the full tape reel on the left spindle and the empty reel on the right spindle.

D. Unwind about 18 inches of tape. The tape should unwind counterclockwise, with the shiny side of the tape facing you.

E. Thread the tape along the tape path to the empty take-up reel. Hold the end of the tape in your left hand. With your right hand, turn the take-up reel counterclockwise two or three turns, so that the excess tape winds securely around the reel.

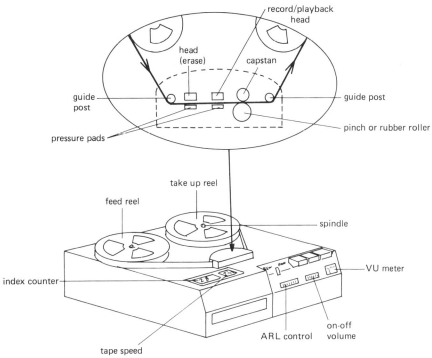

Figure 18-3 *Parts of a reel-to-reel recorder/player—front view.*

Figure 18-4 *Parts of a real-to-reel recorder/player—back view.*

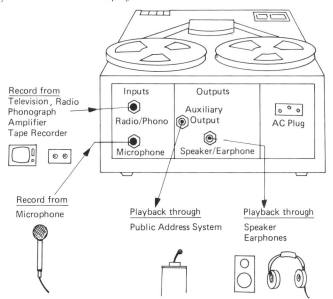

F. Turn the machine ON.
G. Press PLAY or FORWARD.
H. Adjust the tape speed to match the original recording speed.
I. Adjust VOLUME and TONE settings where available.
J. If you are not using AUTOREVERSE (see Glossary), after hearing tape in one direction ("side one"), press STOP. If you need to hear tape in the other direction ("side two"), take the full reel off the right spindle and place it on the left spindle so that the tape unwinds counterclockwise.
K. When you're finished hearing the tape, press STOP. If you have played only side 1, press REWIND to return the tape to the original reel. If you have heard side 2, the tape is now back on the original reel. Remove the tape from the machine.
L. Replace the spindle protectors, if there are any, and turn the machine OFF.

II. **Recording procedures**
A. Plug the cord into an AC outlet.
B. Turn the OFF/ON switch to ON. Let the machine warm up, if appropriate.
C. Place the full reel of tape on the left, and the empty reel on the right spindle. Unwind about 18 inches of tape and thread tape from the full reel to the take-up reel, following the path on the diagram. The tape unwinds counterclockwise. Rotate the tape around the take-up reel a couple of times so that the tape is securely held in place.
D. Set the speed selector switch (see InDepth III).
E. Set the index counter at "000."
F. Plug the microphone into the INPUT or MIC jack and set the mike at the proper distance from the sound input.
G. Activate the mechanism for recording (see InDepth IV).
H. Determine the recording level (see InDepth V). Play back some test comments. If you are not satisfied, make the necessary adjustments.
I. When you are satisfied with the recording level, press STOP. Rewind the tape to index counter "000." You are now ready to record. Activate the recording mechanism. (For AUTOREVERSE, see Glossary.)
J. Press STOP when you are finished.
K. Rewind the tape to "000."

III. **Setting machine speed**
A. *Change the speed when the power is on and the reels are turning.*
B. The faster the tape moves, the higher the quality of the recording:
-7½ ips (inches per second) - for high-fidelity music, broadcast-quality tape recording.
-3¾ ips - good all-around speed for speech and music.
-1⅞ ips - for speech only.

IV. **Activating recording mechanism**
Activate recording mechanism as indi-

cated below, or as indicated for your equipment.
A. Press RECORD lever and hold it firmly as you press PLAY, or
B. Release the RECORD lever or button and press RECORD without pressing PLAY.

V. **Setting recording levels**

Some reel-to-reel recorders have both an automatic and a manual setting for determining the recording level. When using the automatic setting, the machine adjusts the level according to the loudest sound. With the manual setting, the user has flexibility in making the recording level adjustment.

A. Automatic setting.
Set the switch on the machine to AUTOMATIC. You do nothing more. The recorder will adjust the recording level automatically, according to the loudest sound.
B. Manual setting.
Turn the AUTOMATIC setting off, where there is one.
If you are using a remote microphone and it has an ON/OFF switch, turn the switch ON.
Speak into the microphone, and adjust both the recording level knob and the microphone-to-sound input distance until the VU meter needle or LED reading does not peak in the danger zone.
Play back the test comments to check the recording level.

VI. **Tapes**

Buy high-quality tapes. Tapes come in different thicknesses. One mil is for common use. One and a half mil is thicker and stronger. Half a mil is not recommended because it is too thin. The most common reel sizes are 5-inch, 7-inch, and 10-inch reels. Make sure the size reel you buy will fit on your machine.

VII. **See InDepth, Cassette Units** for the following:

Some Special Features

Erasing Procedures

Cleaning and Demagnetizing Heads

Duplicating Tapes

TABLE 18–1

Approximate reel running times in one direction according to tape length and tape recorder speed.			
Tape length (in feet)	Tape recorder speed (ips = inches per second)		
	1⅞	3¾	7½
300	30 min.	15 min.	8 min.
600	1 hr.	30 min.	15 min.
900	1½ hrs.	45 min.	23 min.
1200	2 hrs.	1 hr.	30 min.
1800	3 hrs.	1½ hrs.	45 min.
2400	4 hrs.	2 hrs.	1 hr.
3600	6 hrs.	3 hrs.	1½ hrs.

GLOSSARY

ALC Automatic level control. Automatically sets the recording level according to the loudest sound. Same as *ARL*.

ARL Automatic recording level. Same as *ALC*.

Autoreverse Machine control that allows the user to program ahead of time whether the tape is to run in one direction, then stop; run in two directions, then stop; or to run continuously. The tape never needs to be manually flipped over to change direction. This feature can be used in both playback and recording. It is available for both cassette and reel-to-reel units.

Bias A very high frequency equipment signal emitted during recording that boosts the magnetic pattern being received by the magnetic tape from the recorder. While bias is present in all recorders, some machines have a bias switch. Tape manufacturers may designate at which bias signal their tape works best.

Capstan roller Metal roller that pulls the tape through the machine at a constant, even speed. Must be cleaned periodically.

Cue Forward button that moves the tape more slowly than FAST FORWARD. Allows you to look for a given spot on the tape more easily. Sometimes, where there is no FAST FORWARD button, the term is used interchangeably with CUE.

Dolby Noise reduction system available on some equipment. When recording, it boosts high frequency sounds to cover up the natural hissing sound of the tape. On playback, it cuts back high frequencies boosted during recording. Use Dolby both when recording and on playback, if it is available. A tape recorded with Dolby B should be replayed on Dolby B units. The same applies for Dolby C. Do not use Dolby switch on playback if it was not used when recording because it will mute high frequencies. You can play back a Dolby recorded tape on a non-Dolby machine if you manually decrease treble control.

Heads Units in the recorder or player that either record, play back, or erase information on the tape. Heads must be cleaned periodically to maintain proper sound performance.

Idler roller Roller that presses the tape against the turning metal capstan roller. Needs periodic cleaning. Does not operate during FAST FORWARD and REWIND.

Index counter Usually on the top of the machine. Keeps count of the tape as it passes through the machine. Use it to locate particular passages on the tape. Set to "000" when beginning recording or playback.

Jacks Holes, usually on the sides or the back of the machine, that allow connections to be made either to transmit sounds into the machine, as with a MIC (microphone) or INPUT jack, or to allow sounds to come out of the machine, as with an EARPHONE or OUTPUT jack (see Chapter 22).

LED Light-emitting diode. LED is a visual signal used to indicate sound level at a given moment. It replaces the VU needle meter in some units.

Mode Synonym for function or operation such as STOP mode, REWIND mode, and so on.

Patchcord Connecting cord between two machines (see chapter on Connectors, Jacks, and Plugs).

Pinch Roller See Idler roller.

Plug Element that fits into a jack to complete a connection.

Pulse Audible or inaudible signal placed on a tape, usually at specified intervals, for the purpose of controlling the functioning of one or more other machines, including the unit that is playing the tape.

Remote microphone Microphone that is not built into the machine. It is plugged into the tape recorder and is often hand-held or otherwise held away from the machine.

Rubber roller See Idler roller.

Stereo A concept which attempts to make sound more true-to-life. It is based on the idea that we naturally hear sound coming from more than one direction. In recording, more than one mike is used, placed apart from each other. Sound is recorded and played back on more than one track. More than one speaker is used to play back sound.

Sync Synchronization or working together of two or more machines, such as a tape recorder and a slide projector. This is sometimes accomplished through the use of a pulsed tape (see Pulse).

Track Section or channel along the length of a tape on which sound is recorded or played back. The number of tracks on a tape depends on the capability of the recorder. For example, a one-track machine uses the entire width of the tape for one direction. The tape cannot be used in two directions. A half-track monaural machine records half the tape in one direction and the other half in the opposite direction. With a four-track stereo machine, the same width tape has four narrow bands or channels, two for one direction and two for the other direction.

VU meter Sound level meter for recording and playback. Usually found on recorders on which the recording level can be set manually. Do not allow the meter needle to peak into the red zone (to the far right).

19 Phonographs

CONTENTS

Operating Tips
  Machine 186
  Record 187
  Needle 187

Problems 187

InDepth
  Procedure for Operating a
    Phonograph 189
  Cartridges 189
  Needles 190

Glossary 190

Phonographs give many hours of record-playing pleasure if you recognize that they are delicate sound machines that require careful handling. They come in a wide price range according to such factors as quality of components, whether they are stereo or mono, and whether they have manual or automatic features. They may have three speeds: 33 rpm (revolutions per minute) for long-playing records, 45 rpm for records lasting three or four minutes, and 78 rpm for even shorter recordings. Most phonographs have jacks for connections to PA systems, additional speakers, tape recorders, and headsets for individual listening.

Here is how conventional phonographs work: When the recording is produced, microscopic wave patterns of varying size are cut as record grooves to mechanically reproduce high and low sound patterns. On playback, the phonograph needle vibrates as it follows and faithfully traces the groove patterns. The cartridge which contains the needle converts these vibrations into tiny electrical impulses. The amplifier boosts the minute signals, and the speaker converts the boosted electrical impulses into sound.

OPERATING TIPS

I. Machine tips

   A. Select and set the proper speed. Record speed is indicated on each record label.

   B. When not in use, close the phonograph lid.

   C. The tone arm should be fastened to the arm rest before moving the machine, to avoid damage to the needle and cartridge.

D. Keep the phonograph away from heat.
E. Never touch the tone arm while the automatic changer is in operation.

II. **Record tips**

A. Hold records so that your fingers touch only the edges and the label. Store them in the record jackets.
B. The wrong needle and/or dirt can scratch records. Heat can warp or melt records.
C. Keep records clean. Use a commercially-prepared solution, or prepare your own solution by greatly diluting very mild liquid dishwashing detergent in lukewarm water. Wipe records gently with a soft cloth dampened in the solution. Rinse the cloth and remove the soap from the records. Wipe records dry with a powder puff reserved for this purpose.
D. Store records vertically rather than horizontally; otherwise, they can warp.
E. When stored at an angle for a long time, records will warp.
F. Warped records cannot be repaired.

G. If your phonograph is labeled STEREO/MONO COMPATIBLE, it can play both types of recordings.

Stereo records cannot be played on mono equipment of pre-1964 manufacture without damaging the records and/or the needle. However, mono LPs (but not 45s) can be played on stereo equipment.

III. **Needle tips**

A. Do not allow the needle to bounce on the turntable when the phonograph is being moved (see Operating Tips I).
B. Do not use records that are severely scratched; otherwise, the needle will be damaged. If the records are not in good condition, consider recording them on tape before they get worse (see chapter on Connectors, Jacks, and Plugs for procedures).
C. The needle should always be free of dust. Remove such dust with a soft brush available at your stereo dealer. Do not use your fingers.
D. Where applicable, check that the needle is set for the correct record speed.

PROBLEMS

PROBLEM	CAUSE	SOLUTION
1. No power; power intermittent.	Broken cord.	Repair or replace cord.
2. No sound; sound intermittent.	Detachable speakers unplugged or plugged in incorrectly.	Check speaker connections.
3. No sound or low volume.	Old cartridge.	Replace cartridge.
	No needle in cartridge.	Replace needle.

PROBLEM	CAUSE	SOLUTION
	Volume level low.	Raise volume level.
4. Distorted sound.	Warped record	Do not use record. Replace if necessary.
	Cartridge may be cracked.	Check cartridge. Replace if possible.
	Wrong needle.	Replace needle.
	Defective machine.	Take in for repair.
5. Dull sound, emphasizing bass.	Worn-out needle.	Replace needle.
6. Record has audible static.	Static can occur when removing record from record jacket.	Clean the record (see Operating Tips II).
7. No audible sound distortion but needle scratching record.	Cartridge placed incorrectly in tone arm.	Place cartridge in notched groove. Do not force.
8. Needle repeats or skips at certain spots.	Wrong type of needle.	Check speed and type of record for correct needle.
	Warped record.	Do not use record. Replace if possible.
	Crooked turntable.	Take in for repair.
	Turntable not level on table.	Level turntable, using built-in adjustable feet, or notebook.
9. Needle slides across record.	Needle worn out.	Replace needle.
	Dust buildup around needle.	Remove dust, using soft brush made for needles.
	Needle not contacting record properly.	Check position of cartridge holder in tone arm.
	Wrong type of needle.	Check speed and type of record for correct needle.
	Pickup arm is warped, bent, or detached.	Take in for repair.
10. Cartridge and arm slide across record.	No needle in cartridge.	Replace needle or entire cartridge.

IN-DEPTH

(See Figure 19-1.)

I. **Procedure for operating a phonograph**

   A. Raise the phonograph lid if there is one.
   B. Place the record on the turntable.
   C. Select and set the proper record speed on the machine. Record speed is indicated on each record label.
   D. Where necessary, flip the switch on the tone arm to the appropriate needle. This will match the record speed.
   E. Turn the phonograph ON.
   F. Place the tone arm on the record or activate the automatic function.
   G. When you hear the first sounds on the recording, adjust the tone controls to suit you.
   H. After the record has stopped, remove the record, close the lid if there is one, and turn the phonograph OFF where necessary.

II. **Cartridges**

Cartridges come with needles. The needle is sometimes part of the unit, or it can be replaced without replacing the cartridge. Know either the numbers of your cartridge and needle, or the make and model of your phonograph. When using a cartridge-needle combination, cartridges should be replaced annually.

The weight of your cartridge must be appropriate for your tone arm. However, where possible, a lighter rather than heavier cartridge is preferred because the needle will be more responsive and do less damage when traveling through the record grooves. Consult your dealer.

Figure 19-1 *Parts of record player with built-in amplifier, as indicated by presence of volume controls. Amplifier can be separate. Speaker can be internal or external.*

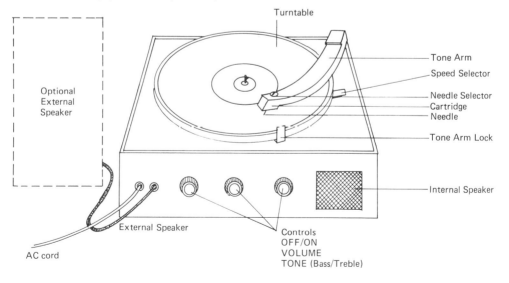

Magnetic cartridges are used primarily in higher-quality equipment. Ceramic cartridges are used in less expensive equipment. They are not usually interchangeable.

II. Needles

   A. Materials used in making needles are metal, sapphire, and diamond. Diamond is best.

   B. On many phonographs, one side of the needle will have a 78 rpm needle; the other, one for both 33 1/3 and 45 rpm.

   Needles must be matched to the width of the grooves on the record. This varies according to the record speed and whether it is a stereo or a mono recording. Using a needle at the wrong speed will prevent maximum sound reproduction.

   33 1/3 / 45 rpm mono needle is NOT interchangeable with the 33 1/3 / 45 rpm stereo needle. However, phonographs made since 1964 are mono/stereo compatible and are labeled as such.

   C. When using non-diamond needles, or playing older or scratched records, needles should be changed at least once annually. Otherwise, the needle will begin to flatten and skip across the record. If you have a filmstrip projector/phonograph combination, the flat needle will result in false cuing of the automatically advanced filmstrips.

GLOSSARY

Digital recording An advanced method of making and playing back records. The original sounds that come from the microphone are turned into number-coded electrical impulses according to their pitch and loudness of the sounds. On playback, a laser beam scans the record for the digital pulse marks. There is no needle and there are no record grooves. Since the information is stored beneath the top disc layer, dirt or fingerprints cannot affect sound reproduction.

Linear tracking A way in which some tone arms move across the turntable. The tone arm is that part of the turntable that at one end holds the cartridge/stylus and takes it across the record. Conventionally, a tone arm swings onto the record and is delicately balanced at the non-cartridge end. In linear tracking, the non-cartridge end of the tone arm is firmly attached to and operates along a moving track that goes from right to left and back. Because the tone arm moves horizontally across the record, tracking is consistent, with less record damage and better sound reproduction.

Stroboscope A simple, inexpensive device used to check the accuracy of the speeds on your phonograph. The concept may also be built into your phonograph.

Tweeter A type of speaker that reproduces high-frequency sounds. A sound system can have several loudspeakers built into a speaker box, each reproducing certain sound frequencies. For example, there could be a tweeter, woofer and a speaker for middle-range sounds.

Woofer A type of speaker that reproduces low-frequency sounds.

20 Microphones

CONTENTS

Operating Tips
    General Tips 191
    Establishing Sound Levels 192
    Microphone Placement 192
    Microphone Handling 192
    Eliminating Undesirable Sounds
        During Recording 192

Problems 193

InDepth
    Types of Microphones 195
    Microphone Maintenance 195

Glossary 195

A microphone (mike or mic for short) is an instrument used to convert sounds into electrical current. Microphones are used with many types of equipment, including audio and video sound equipment and public address systems. Note the following:

1. The microphone must be electrically compatible (have correct impedance) with the equipment you are using.

2. The connections between the microphone and other equipment must be physically compatible. In other words, the microphone plug must fit into the equipment jack.

When requesting a microphone, specify the make and model of the machine you are using.

OPERATING TIPS

I. General tips

    A. If the equipment has both a built-in microphone and a microphone jack for an added microphone, use of the added microphone will override the built-in microphone. You may use a separate microphone when your built-in mike is defective.

    B. Always connect the microphones to the MIC input jacks.

    C. Make sure the cord connections are not frayed or broken.

    D. Observe the correct polarity when inserting the battery for a condenser microphone. If the polarity is reversed, the microphone will be ruined.

E. See also Operating Tips III, Tape Recorders chapter.

II. Establishing sound levels

Sound levels will differ from room to room, indoors and outdoors, as well as from voice to voice. Always test the level before using the microphone for your presentation.

A. For recording:
   1. Speak normally into the microphone.
   2. Adjust the VOLUME knob. Watch the recording level meter, if available. The meter should not peak into the red area. If you have no meter, speak into the microphone while recording, then check the recording level on playback.

   Without a meter, chances are your machine sets the recording level automatically. With recording levels set automatically, the loudest sound, whether background noise or speaker's voice, will control the recording level. Automatic recording levels will be labeled ALC or ARL on some machines. In this case, the VOLUME knob is for the playback level only.

B. For mixers and PA systems, see In-Depth III, PA Systems chapter.

III. Microphone placement

A. Place the microphone near you. Hold the mike 3 inches slightly to the side of, and 6 to 10 inches away from, your mouth; otherwise, "b" and "p" sounds will pop. Speak in a normal voice.

B. In a very noisy room, keep the microphone close to your mouth, but speak lower than usual. Your breath should not be heard.

C. If you're using a separate rather than a built-in microphone, place it far enough away from the equipment to minimize motor sounds.

D. Do not vary the distance between the mike and the person talking into it once you have determined the proper distance.

E. For use of the microphone by more than one person, all subjects should be at about an equal distance from the microphone if each is speaking at equal volume. Use an omnidirectional microphone designed to pick up sounds from all directions. Handling noises will be heard if the microphone is passed from person to person.

IV. Microphone handling

A. Do not shuffle papers or drum on the table with fingertips. Don't blow or whistle into, or tap on the microphone, as this can damage it.

B. Where it is possible to increase the volume while using the microphone, increase it slowly so that it will appear natural.

C. In recording, additional microphones may be used if sufficient jacks are available in the recorder. In most cases, a mixer is preferable (see Glossary).

V. Eliminating undesirable sounds during recording

A. Record in a fairly noiseless room. The following environment noises may be picked up by the microphone:

- Heat and ventilation sounds
- Fluorescent hum
- Outside noises

B. Record during quiet times, such as late afternoon or evening.
C. Use a room that has carpeting.
D. Reposition the microphone away from unwanted sounds.
E. Use a unidirectional microphone (see Glossary).

PROBLEMS

(See also Problems for chapters on equipment to which a microphone can be attached.)

PROBLEM	CAUSE	SOLUTION
1. No sound.	Defective microphone.	Replace microphone.
	Poor connection.	Check cord connections at microphone and at equipment jack.
	Microphone switch OFF.	Turn switch ON.
	Microphone not suited for machine.	Request another microphone. Know make and model of equipment you are using.
	Condenser microphone battery old or missing.*	Replace microphone battery.
2. Feedback.	Volume level too high on recorder, PA, etc.	Lower volume level.
	Microphone too close to loudspeaker.	Keep microphone behind loudspeaker.
	Talking too loudly into microphone.	Speak in a normal voice.
	Microphone too close to performer's mouth.	Keep microphone 6 to 12 inches away.
	Microphone cable too close to electrical extension cord.	Separate cords.

*A hand-held condenser microphone has batteries. You can tell if you have such a microphone if the barrel unscrews from the head. The battery compartment is inside the microphone barrel.

PROBLEM	CAUSE	SOLUTION
	Recording with remote microphone in tape recorder; MONITOR or SPEAKER switch ON without using headset.	Turn MONITOR or SPEAKER switch OFF while recording, unless using headset.
3. Distorted sound.	Talking too loudly into microphone.	Speak in a normal voice.
	Microphone not suited for machine.	Request another microphone. Know make and model of equipment you are using.
	Condenser microphone battery old or weak.*	Replace microphone battery.
4. Garbled sound.	Defective tape.	Replace tape. Record again.
	Playback on reel-to-reel tape recorder at wrong speed.	Play back on same speed as recorded.
	Condenser microphone battery old or weak.*	Replace microphone battery.
5. Extraneous noises in recording.	Too close to outside noises, such as heating and vent noises.	See Operating Tips V.
	Reversed polarity on batteries for condenser microphone.	Microphone ruined. Replace.
6. Weak sound.	Defective microphone.	Replace microphone.
	Microphone too far from performer.	Bring microphone closer to performer.
	Volume level too low.	Raise volume level.
	Microphone not suited for machine.	Request another microphone. Know make and model of equipment you are using.
	Condenser microphone battery old or weak.*	Replace microphone battery.

---

*A hand-held condenser microphone has batteries. You can tell if you have such a microphone if the barrel unscrews from the head. The battery compartment is inside the microphone barrel.

IN-DEPTH

I. Types of microphones

A microphone is to its equipment what a lens is to a camera. You may need different microphones for different purposes. Microphones are described according to:

A. Pickup patterns, the direction from which the mikes are designed to accept sound (see Glossary, Unidirectional, Bidirectional, and Omnidirectional Microphones).

B. The type of element in the microphones used to change sound into electrical impulses (ceramic or crystal, dynamic or moving coil, condenser, ribbon). Condenser microphones, most of which contain batteries, and dynamic microphones reproduce sound very accurately.

C. The range of sounds (frequency), from highest to lowest, at which mikes are effective.

D. The physical way in which they are handled, such as on a tabletop, hanging around the neck, and so on.

You will need this information only if you are doing projects that have special sound requirements. As mentioned earlier, noting the make and model of your equipment will provide a microphone to suit general needs.

II. Microphone maintenance

A. Make sure the cord connections are not frayed or broken.

B. Make sure the microphone is not handled roughly or dropped.

C. To replace the plug, see InDepth V, Connectors, Jacks, and Plugs chapter.

GLOSSARY

ALC Automatic sound level control for recording. Same as ARL.

ARL Automatic recording level. Same as ALC.

Bidirectional microphone Microphone that picks up sounds from two opposite directions, such as in an interviewing situation.

Cardiod microphone See Unidirectional microphone.

Feedback Undesirable high-pitched squeal caused by improper positioning of microphone or auxiliary speakers.

Impedance Electrical term meaning the opposition that a circuit presents to the flow of electricity. Equipment and microphones are classified as having low or high impedance. The impedance of the two must match. Impedance is measured in ohms.

Mixer Simple box used to feed several microphones or other inputs into one machine. With the mixer, the volume for each sound source can be determined independently without affecting the other inputs. In addition to independent volume controls, some mixers have a master volume control that adjusts the volume of all of the inputs after each has been set. Microphone mixers are available as AC-operated or battery-operated units (see Figure 20–1).

Omnidirectional microphone Microphone that picks up sounds from all directions. Tape recorders most commonly come with such mikes.

Unidirectional microphone Microphone that picks up sound from only one direction. It is used to keep out unwanted background sounds while you're talking into the microphone. This is an inexpensive mike, available at audio stores.

196 *Microphones*

To determine if you have one, establish the proper volume for when the mike is pointed at your mouth and you are talking into it. Then, face the head of the mike away from you at a 90-degree angle. If the sound drops off dramatically when you speak into it, it is a unidirectional mike. This is also called a cardiod microphone.

Windsock Inexpensive foam rubber jacket used to cover pickup end of microphone. Available for most microphones today. Useful for eliminating whistling wind noise.

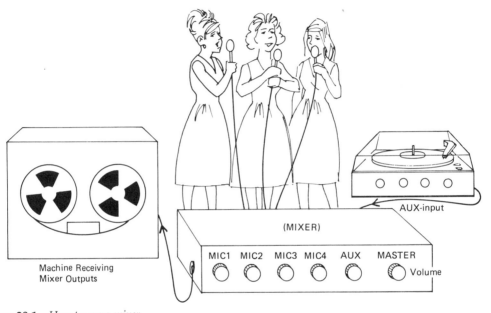

Figure 20-1 *How to use a mixer.*

21 Headsets, Jackboxes, Listening Stations

CONTENTS

Operating Tips
    Headsets 198
    Jackboxes and Listening Stations . . 199
Problems
    Individual Headsets 199
    Jackboxes and Listening Stations . . 200

InDepth
    Plugs for Headsets. 200
    Procedure for Using Jackboxes and
        Listening Stations 201

Glossary 201

HEADSETS

Headsets provide a listening opportunity for one individual on a particular machine without disrupting other activities in the room. Most machines equipped for sound have output jacks for headsets. Radios and televisions either have such jacks or can have jacks built into them. The headsets must be electrically compatible (match impedance) with the equipment with which they are being used. This is usually not a problem.

    Headsets or headphones come in a great variety of makes, styles, and prices. Cheap ones cost as little as $7 to $10, while more expensive ones cost $30 or more a piece. It is recommended that you purchase expensive ones for home use. For organizational or school use, buy inexpensive ones and discard them when they no longer work.

    Storage for several headsets comes in various shapes, in bags, or hanging from a bracket. Choose what best suits your space needs.

JACKBOXES

A jackbox allows several individuals to wear headsets at the same time. In this way, they can all hear the same selection at the same time from one piece of equipment without disturbing a larger group.

    A jackbox is a small box with several jacks (usually eight to ten) and a plug to connect the box to the output jack of the originating machine. Jackboxes usually have quarter-inch jacks.

LISTENING STATIONS

A listening station is a jackbox with some type of storage space for the headsets, or the complete hookup with jackbox, headsets, originating machine, and the physical location where it is all set up. The more expensive listening stations have a volume control for each jack. Since individuals in a group may be using headsets of different makes that transmit different ranges of volume, these separate volume controls are useful.

There are listening stations with no jacks and accompanying headsets with no plugs. Jacks and plugs have been replaced by magnetic couplers, thus eliminating many problems. Volume controls are available on each headset (see Figure 21–1).

OPERATING TIPS

I. Headsets

A. Plug a headset into the OUTPUT, HEADPHONE, or EARPHONE jack of your equipment.

B. To use several headsets for listening to one piece of equipment, use a jackbox or listening station.

C. When finished, carefully remove the headset plug from the jack of the machine. Do not pull by the cord.

D. On more expensive headsets, the cushions can be removed from the headset, washed in soapy water, rinsed, and dried.

E. Headset cushions are often replaceable.

F. Some headsets have volume control on them, instead of requiring you to use the volume control on the machine or jackbox.

G. Stereo headsets can be purchased for listening to stereo equipment.

Figure 21-1 *How about magnets to connect headsets! (Courtesy Murdock Corporation).*

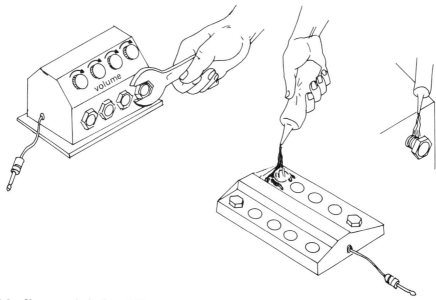

Figure 21-2 *Keep your jacks from falling in the jack box.*

II. Jackboxes and listening stations

(See Figure 21-2.)

Tighten the nut around each jack when you first get the jackbox, and check it periodically thereafter. Use nail polish or glue to seal the threads so that the nut cannot be removed, and the jack will not fall into the box.

PROBLEMS

PROBLEM	CAUSE	SOLUTION
Individual Headsets		
1. Broken cords.	Rough handling; pulling headset out of machine jack by cord instead of by plug.	If your headset has a molded plastic plug, cut it off. Trim cord to past where it is broken. Replace plug with a plug having screw connections, or discard headset (see InDepth V, Connectors, Jacks, and Plugs chapter).
2. Inside of headset falls out.	Headset fell or received a blow.	Repair or discard headset.
3. No sound, weak sound.	Defective or inappropriate headset.	Replace headset.

200 *Headsets, Jackboxes, Listening Stations*

PROBLEM	CAUSE	SOLUTION
Individual Headsets		
	Headset plug in wrong output jack of machine.	Try other output jacks.
	Defective equipment.	Check Problems section in chapter on equipment you are using.
	Volume turned down.	Adjust volume.
Jackboxes and Listening Stations		
4. Jack mounting falls into box.	Nut around jack falls off.	Tighten nut with pliers; seal threads with glue or nail polish (see Fig. 21–2).
5. Weak sound when using two jackboxes.	Too many headsets.	Reduce number of headsets.
6. No sound.	Defective headset.	Replace headset.
	Jackbox cord in wrong output jack of machine.	Try other output jacks.
	Defective jackbox cord.	Replace cord.
	Volume turned down.	Adjust volume.

IN-DEPTH

1. Plugs for headsets

Many headsets have quarter-inch plugs because many machine output jacks are a quarter-inch. Miniature jacks, particularly for EARPHONE jacks on some cassette recorders, require adaptors for the headset to fit. These are very inexpensive. Buy adaptors with a flexible cord between the plug and jack (see Figure 21–3).

Figure 21-3

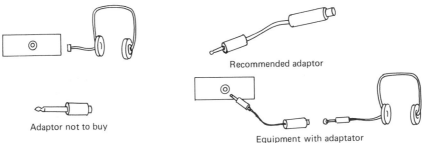

Select a flexible-cord adaptor

II. Procedure for using jackboxes and listening stations

1. Insert the headset plug(s) into the jackbox.
2. Connect the appropriate cord from the jackbox to the HEADPHONE jack of the machine that is providing sound.
3. Turn the machine on and play.
4. Adjust the volume on the machine and/or the headsets, where possible, or the jackbox.

GLOSSARY

Jack Hole or holes in equipment, usually provided on the side or the back, designed to accept a connector from another piece of equipment. Depending on whether it is an output or input jack, it emits or receives a signal.

Plug Connector at the end of a cord that is inserted into a jack to complete the electrical circuit for emitting or receiving a signal.

22 Connectors, Jacks, and Plugs

CONTENTS

Operating Tips 202

Problems 203

InDepth
Making Proper Connections 204

Procedure for Transferring
    Formats Using Patchcord 204
Setting Sound Level 206
Plug Types 207
Replacing a Plug on
    a Patchcord 208

Connecting two pieces of equipment, or patching, usually requires a connector or patchcord. This is a cord or wire with a plug at either end.

Machines that receive and/or disseminate sound have holes called jacks. The electrical circuit is completed when a plug is inserted carrying sound to or from the other instrument.

Know the make and model of your equipment to determine the proper plugs for the patchcord. Plugs need to match the size, shape, AND wiring of the jacks in order to fit into the jack or work properly. Plan your connections ahead of time.

Patchcords can be used to transfer sound formats—that is, to change the medium through which a program is presented. Here are some examples of format changes: reel-to-reel/cassette, record/tape, PA/tape, sound format to super 8 sound or videotape.

Transferring formats can be done without a patchcord by having two pieces of equipment near each other. You play the original program on one machine, and record it onto the new format machine, using a microphone attached to the new format machine. However, with this method, noises such as air conditioning and people talking may be recorded onto the new format. On the other hand, the patchcord or connector will transfer sound directly from one machine to another without exposure to extraneous sounds. Both pieces of equipment must be compatible. If you have doubts, check with your audio/visual dealer.

OPERATING TIPS

1. Make sure the plug is pushed all the way into the jack.

Connectors, Jacks, and Plugs 203

2. Never remove a plug from a jack by pulling the cord. Hold onto the plug itself and remove.

3. Do not connect outputs to outputs or inputs to inputs. This could seriously damage machines.

4. Always connect microphones to MIC input jacks (see also Operating Tips III, Tape Recorders chapter).

5. Always connect loudspeakers to speaker or headphone jacks.

6. If you do not obtain any sound or the desired quality of sound after using what you think is the appropriate plug, try another plug before suspecting a malfunctioning machine or defective sound source.

7. You may often wonder how to connect two machines to each other. Check Table 22–1 for connection possibilities.

PROBLEMS

PROBLEM	CAUSE	SOLUTION
Patchcord Problems in Transferring Formats		
Try solutions below, then record again. Output or originating machine is machine A. Input or take-up machine is machine B.		
1. No sound on machine A or no sound on machine B when tested independently.	Defective AC outlet.	Try another AC outlet.
	Defective record or tape.	Use another tape.
	Tape on wrong side.	Rewind tape and turn over.
	One or both machines defective.	Replace machines.
	See also Problems 1, 2, 3, 14, Tape Recorders.	
2. No playback sound on machine B, although machines A and B were each tested before using.	Patchcord only plugged into machine A.	Plug other end into machine B.
	Plugs are in wrong jacks, such as input instead of output or vice versa.	Check jack connections.
	Patchcord and/or plugs and/or machine jacks defective.	Replace patchcord. If defective jack, replace machine.

Connectors, Jacks, and Plugs

PROBLEM	CAUSE	SOLUTION
Patchcord Problems in Transferring Formats		
	Patchcord plugs and machine jacks electrically incompatible, even when plug fits jack.	Make and model of equipment determines kind of patchcord plug.
	Jack or plug corroded.	Clean plug with very fine steel wool. Rotate plug in jack five or six times.
		Take in for repair.
3. Interference in recording when using patchcord.	Machine picking up extraneous signals such as radio, CB, or stereo.	Move to new location. Try other machines.
		Use shorter cord, if available.
		Order patchcord with shielded wire.
4. Sound distorted.	Patchcord plugged into MIC input.	Change plug to AUXILIARY INPUT.
5. Intermittent sound.	Plugs not pushed all the way into jacks.	Push plugs as far as they will go into jacks.
	Loose wire in one or both plugs.	If possible, unscrew plug covers. Check that patchcord wires are tightly connected to each plug.

IN-DEPTH

I. Making the proper connections

(See Table 22–1.) To use this chart:

A. Look at your machines.

B. On the chart, find the jack name for machine A that best corresponds with the jack name for machine B to connect *your* equipment.

C. Follow the directions in InDepth II.

II. Procedure for transferring formats using patchcord

A. For some types of equipment, try to obtain machine B with a sound level meter.

B. Check out sound for each machine separately by playing back the original program on machine A and, where possible, recording and playing back a tape on machine B.

Figure 22–1
*Making the proper sound connections**

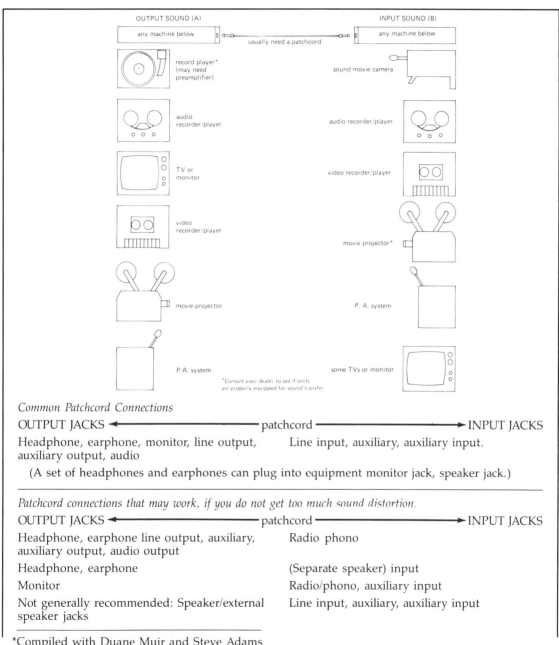

Common Patchcord Connections

OUTPUT JACKS ←─────────── patchcord ───────────→ INPUT JACKS

| Headphone, earphone, monitor, line output, auxiliary output, audio | Line input, auxiliary, auxiliary input. |

(A set of headphones and earphones can plug into equipment monitor jack, speaker jack.)

Patchcord connections that may work, if you do not get too much sound distortion.

OUTPUT JACKS ←─────────── patchcord ───────────→ INPUT JACKS

Headphone, earphone line output, auxiliary, auxiliary output, audio output	Radio phono
Headphone, earphone	(Separate speaker) input
Monitor	Radio/phono, auxiliary input
Not generally recommended: Speaker/external speaker jacks	Line input, auxiliary, auxiliary input

*Compiled with Duane Muir and Steve Adams

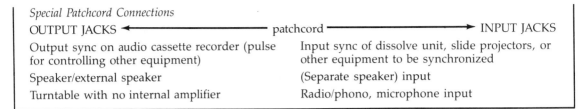

Figure 22-1 *(continued)*

C. Connect the patchcord to the appropriate input and output jacks.

D. Set the volume at approximately the same level for both machine A and machine B (see InDepth III). One loud and one soft volume distorts the sound.
Do not turn the volume over halfway on either machine unless necessary. For those machines that have meters, do not let the needle peak (go into the red area of the meter).

E. Where possible, before making the actual recording, have a trial run to test out the equipment using the patchcord. Adapt the tape recorder trial run directions below for the type of equipment you are using:

1. Insert program in machine A.
2. Insert blank tape in machine B.
3. Press RECORD and PLAY on machine B.
4. Press PLAY on machine A.
5. Make sound level adjustments on each machine. (See InDepth III below.)
6. Stop machines. Rewind.
7. Play back new tape on machine B. Control playback volume by raising or lowering volume level on machine B.

You will not get better sound quality on the new tape than what was on the original.

F. Rewind to the beginning. You are now ready to transfer formats.

G. Repeat step E above for actual transfer.

III. Setting sound level for machine B when transferring format

A. Rule of thumb: When connecting two pieces of equipment, initially set volume controls halfway, where possible.
Machine A—when using AUXILIARY (LINE) OUTPUT for your patchcord connection to machine B, the VOLUME knob will be ineffective! The machine will determine the volume output.
Machine B—manual versus automatic recording level. You can set the recording level manually when the machine offers a labeled choice of MANUAL or ALC (automatic level control). When there is only automatic level control there is probably no sound level meter and the VOLUME knob only functions during playback. ALC produces satisfactory sound level, unless the original sound was too loud, too low or distorted.

B. Trial and error: Connect machine with patchcord, according to In-Depth II above. As you record, make sound level adjustments on machine A, where possible. If machine B has a recording level meter, do not allow needle to peak into red area.

IV. Plug types

(See Figure 22–2)

Some equipment takes quarter-inch plugs (to fit quarter-inch jacks). Other equipment uses mini plugs, submini plugs and din plugs. The most expensive equipment, particularly turntables and stereo tape recorders, often uses phono (RCA) jacks.

V. Replacing a plug on a patchcord

(See Figure 22–3.)

A. Buy a plug with a screw connection (available at audio dealers).

B. Cut off the old plug where it joins the patchcord. (1)

C. Where you have just cut off the old plug, scrape a half-inch of cord, using a knife or blade. (2) This should expose the wires. Slip the plug cover onto the patchcord. (3)

Figure 22-2 *Plug types.*

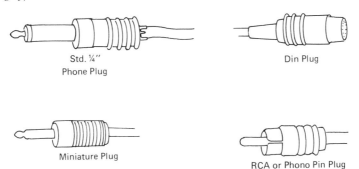

Figure 22-3 *Replacing a patchcord plug. (Some plugs do not have screw terminals. Connections must be soldered.)*

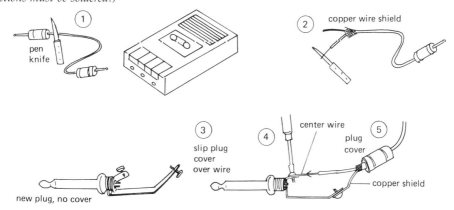

D. Wrap the center, or main, wire of the cord around the new plug screw as shown below. Tighten with a small screwdriver. (4)

E. If you have a copper wire (shield), wrap it around the shield screw. Tighten with a small screwdriver.

F. Slip the plug cover over the connections. (5)

G. You now have a new plug at the end of your patchcord.

23 Public Address Systems

CONTENTS

Operating Tips	209
Problems	210
InDepth	
Basic PA Switches, Controls, Jacks.	211
Hooking Up a PA System	211
Setting Master Volume Control. . . .	212
Speakers.	212
Glossary	213

A public address system (PA) is an amplifier of sound. It is used to communicate with large audiences or audiences that are not in the same location as the performer. A very basic PA system consists of a microphone, amplifier, and loudspeaker. PA systems come in many sizes—examples are the small portable variety, the lecturn-type large PAs used for sports events, and PAs used for communicating throughout a building. Some are battery-operated, some run on either batteries or AC, and some are powered by AC only. The price ranges from a few dollars to several thousand.

OPERATING TIPS

1. PA system should be off when plugging and unplugging microphones and loudspeakers; otherwise, damage may occur.

2. If there are two or more microphone inputs (MICROPHONE 1, MICROPHONE 2, or VOLUME 1, VOLUME 2, etc.), keep all volume controls set at "0," except the one in use.

3. Do not bring the microphone too close to your mouth (see also Operating Tips I, III, Microphones).

4. With some battery/AC systems where there is no ON/OFF switch, do not leave the microphone plugged in when not in use, as this activates the batteries.

5. Do not make extraneous noises near the PA microphone, because the microphone will broadcast the sound.

6. Always test the system before using it.

PROBLEMS

PROBLEM	CAUSE	SOLUTION
1. Feedback.		The lower the volume level of the PA, the less chance of feedback.
	Volume too high on master volume, microphone controls, or both.	Set controls carefully (see InDepth III).
	Talking too close to or too loudly into microphone.	Adjust microphone position and volume. Speak in a normal voice.
	Walking in front of PA loudspeaker with microphone hooked into PA.	Stay behind PA lectern.
	Two auxiliary loudspeakers facing each other. Loudspeakers facing PA. Loudspeakers too close to PA.	Reposition auxiliary loudspeakers, pointing them away from microphone.
2. Microphone does not work. Sound not loud enough, even if volume turned up.	Defective microphone. Usually broken wire in cord.	Have extra microphone on hand.
	Microphone not suitable for your PA.	Know make and model of PA when requesting microphone.
	Microphone plugged into AUXILIARY INPUT jack.	Change microphone to MIC INPUT jack.
3. Sound choppy; no sound.	Batteries in PA run down.	Remove batteries immediately and replace.
	No power on AC operation.	Check cord and outlet.
	Defective microphone.	Replace microphone.
	Loudspeaker not plugged in.	Check loudspeaker connections.
	No sound if MASTER VOLUME control, where available, is turned down.	Turn MASTER VOLUME three-quarters up.

	No sound if microphone volume is not turned up.	Adjust sound levels.
4. Distorted sound.	Auxiliary equipment plugged into PA system is not properly connected or is not electrically compatible with PA.	(See Table 22–1.) Check connections; contact dealer.

IN-DEPTH

I. Basic PA switches, controls, jacks

A. Front panel:
1. VOLUME controls for one or more microphones.
2. MASTER VOLUME control, for total output of sound coming from the PA—not found on all units.
3. Light switch for a podium-type PA system.
4. Battery/AC switch—not found on all units.

B. Back panel:
1. MICROPHONE and/or MIC INPUT jack for the master microphone.
2. AUXILIARY MICROPHONE (MIC) jacks to plug in additional microphones.
3. AUXILIARY INPUT jacks to plug in record player, radio, or tape recorder for the PA to project these input sounds.
4. AUXILIARY SPEAKER jacks for extra loudspeakers.

II. Hooking up a PA system

(See Figure 23–1.)

A. Plug AC cord into the AC outlet, or, where available, turn on the battery switch.
B. Keep the ON/OFF switch OFF.

Figure 23-1 *Common public address hookup. Usually, front panel contains volume controls, back panel contains equipment terminals.*

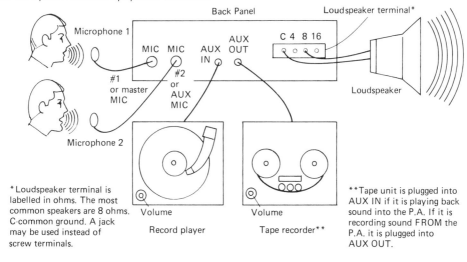

*Loudspeaker terminal is labelled in ohms. The most common speakers are 8 ohms. C-common ground. A jack may be used instead of screw terminals.

**Tape unit is plugged into AUX IN if it is playing back sound into the P.A. If it is recording sound FROM the P.A. it is plugged into AUX OUT.

C. If there are separate loudspeakers, plug them into the loudspeaker outlets.
D. Plug the microphone into the MICROPHONE or MIC INPUT jack.
E. Add an additional microphone if necessary, using additional MICROPHONE inputs.
F. Plug in additional equipment such as record player or tape recorder, using AUXILIARY (LINE) INPUT jacks.
G. Turn all volume controls to "0".
H. Turn the ON/OFF switch ON.
I. Adjust volume (see In-Depth III below).
J. Talk into the microphone, keeping it 6 to 10 inches away from your mouth.
K. When you are finished, turn the ON/OFF switch to OFF before unplugging the microphones and loudspeakers.

III. **Setting master volume control**

There is no sound from the PA system if the MASTER VOLUME control is turned down. PA systems often have such a volume control if the unit provides two or more separate volume controls for microphones and/or auxiliary equipment.

A. Method 1—Lower the PA unit MASTER VOLUME to "0". Raise the individual microphone volumes to their highest levels. Slowly adjust the MASTER VOLUME control until you hear the slightest feedback. Lower the MASTER VOLUME just below feedback level.

B. Method 2—Turn the MASTER VOLUME three-quarters up. Adjust the proper volume for individual units.

Microphones. Adjust each microphone volume on the PA unit. Test each level by speaking in a normal voice and having a listener at the back of the room to gauge the sound.

Auxiliary equipment (record player, radio, tape recorder, etc.). If you are using AUXILIARY (LINE) OUTPUT on the auxiliary equipment, adjust the volume through the PA unit AUXILIARY INPUT VOLUME control.

In other cases, set the auxiliary equipment VOLUME at mid-level setting. Set the PA unit AUXILIARY INPUT VOLUME control at mid-level setting. Test the sound. Raise the auxiliary equipment volume, if necessary. Raise the PA unit auxiliary input volume if you are still not satisfied.

If necessary, adjust the master volume.

IV. **Speakers**

A PA system can have built-in loudspeakers. Whether or not you need additional speakers depends on the size of the room. You need additional amplifiers and/or speakers if you keep having to turn up the PA unit MASTER VOLUME to be heard intelligibly.

Here's how to relate a speaker to your amplifier. A speaker can put out 1½ times the power rating of the amplifier. If an amplifier has a 50-watt rating, the speaker can be rated at 75 watts.

GLOSSARY

Amplifier Component in the PA unit that greatly increases the electrical signal from the microphone and passes it on to the loudspeaker component.

Feedback Undesirable high-pitched squeal caused by improper use of the microphone or improper positioning of extra loudspeakers.

Jacks:

- INPUT jacks are holes in the PA system designed to accept a plug connector from a radio, phonograph, and so on. Once the proper connection has been made, the additional equipment will feed sound into the PA system.

- OUTPUT jacks are holes in the PA system designed to connect equipment for recording from the PA or for additional amplification of the PA.

- AUXILIARY jacks are additional input or output jacks used to connect additional equipment to the PA system.

Loudspeaker Component, either built into or added to the PA, that converts the electrical signal from the amplifier into sound heard by the audience. Also called *speaker.*

Master volume control Control for all PA sound output.

Microphone See Microphones chapter.

Speaker See Loudspeaker.

Tweeter See Phonographs, Glossary.

Woofer See Phonographs, Glossary.

24 Television

CONTENTS

Operating Tips
   General Tips 214
   Using Cable 215
Problems 215
InDepth
   The Fine Tuner 216

   Automatic Fine Tuner (AFT) 217
   Preset Color Level Control
     (Automatic Control for Color) .. 217
Glossary 217

Today millions of Americans have television sets, whether they be large or portable, color or black and white. The models and prices vary greatly. Television watching has had a major impact on our lives, particularly on our educational and entertainment values.

   A revolutionary concept in TV watching is just about here. It is digital TV, based on digital electronic computer technology. Digital TV refers to the way in which the incoming TV signal is processed within the set (see Glossary). Digital TV will improve picture quality and provide new features. Imagine, for example, freeze-framing or zooming in on a picture on your screen, or watching one channel while viewing a different channel in the corner of the screen. Someday a digital TV user will be able to call up on the screen a list of upcoming TV programs, enter his preferences, be advised of programming conflicts and, with one more step, program the VCR!

   Digital technology will soon also be useful in two-way transmissions using the conventional analog TV set. How about shopping electronically? Home security systems are also possible.

   Digital technology has almost unlimited possibilities!

OPERATING TIPS

I. General tips

1. Unless you are skilled, generally make adjustments only on the front of the set. The back is for the repair person.

2. The fine tuner is perhaps the most important feature for the TV user. It is somewhat like tuning the channel selector on a radio to obtain the best station reception.

3. On color sets, your best color picture is obtained when you are satisfied with the flesh tones.

II. Using cable

1. It is illegal to hook into a cable system on your own without making arrangements with the cable company in your community.

2. If you are considering hooking up to cable, you may not have to pay a hookup charge if the cable company is first connecting your neighborhood. This varies with the particular cable company.

3. If you are about to buy a new TV set, ask if a cable ready unit would be useful with your community's cable system set up. If the system uses a converter box, cable ready units in most cases will not be an advantage. Also your TV remote control will not be effective for switching stations, if the cable system is more than a twelve channel system. Now in the developmental stage, work is being done to produce remote control units and converter boxes that will be compatible with each other.

4. If the cable system uses a converter box with your television, you need an additional converter box for your VCR to be able to watch one program and record another. Check with your cable company.

PROBLEMS

PROBLEM	SOLUTION	PROBLEM	SOLUTION
1. No picture.	Check outlet connections. If hooked to cable, call the cable company. On tube-type sets, allow set to warm up one or two minutes. Turn fine tuner in either direction. Check brightness control. Take in for repair. If using monitor TV, selector may be on VTR. Switch selector to TV.	3. No color.	Set is black and white. Program is in black and white. Turn fine tuner in either direction. Check programs on other channels. Release AFT. Balance color controls. Lock AFT.
2. No sound.	Turn the fine tuner in either direction. Turn VOLUME up. Take in for repair.	4. Fuzzy picture.	Turn fine tuner in either direction. Adjust focus (some sets do not have focus controls for user). Turn rabbit ears, if you have them, until you get a good picture. Have repair shop check master antenna system.

PROBLEM	SOLUTION	PROBLEM	SOLUTION
5. Bars tilted left or right; picture tears sideways.	Where possible, adjust HORIZONTAL HOLD. Check for local interference, such as other appliances. If you cannot get rid of bars, take in for repair.	10. Snow with or without picture.	Locate local interference. Turn other appliance off. Check antenna connection. If you are connected to cable and have a converter box, call the cable company. Your set will be reprogrammed within minutes.
6. Vertical roll, upward or downward.	Adjust fine tuning. Where possible, adjust VERTICAL HOLD. Change location or direction of antenna.		
7. Ghosts.	Change location or direction of antenna. If hooked to cable, call the cable company. Adjust the fine tuning.	11. Frequent snapping or popping sound when set is already warmed up.	Turn off, unplug set IMMEDIATELY; take in for repair.
8. Cannot get a particular station.	Check to see that master antenna is attached to wall, or adjust rabbit ears. If hooked to cable, call the cable company. Turn fine tuner in either direction.	12. Screen blank except for very bright line.	Turn off IMMEDIATELY; take in for repair.
		13. White flashes in picture.	Locate interference from major appliance. Turn appliance off.
9. Poor contrast, washed-out picture.	Adjust fine tuning. Adjust BRIGHTNESS. Adjust color control. Adjust contrast control.	14. Buzz or hum in sound.	Locate local interference. Turn appliance off. Adjust fine tuning.

IN-DEPTH

I. **The fine tuner**
   A. Use the fine tuner first when trying to get the best picture, sound, and color.
   B. Turn the tuner to the left or to the right until you have the desired picture quality.
   C. On some sets, if turning the fine tuner does not seem to have any effect, try pushing in or pulling out the knob while turning it.
   D. If you turn the tuner too far in one

direction, it may start slipping. Turn it back the other way.

II. Automatic fine tuner (AFT)

Some sets have an automatic fine-tuning control that locks in the fine-tuning adjustment you have made for each station with the fine tuner, and to the extent possible maintains this picture and sound quality. The AFT can be found on both black and white and color TV sets.

Make adjustments as follows:

A. Turn the set on to a particular station.
B. Release the AFT.
C. Use the fine tuner, not the AFT, to get the most pleasing picture.
D. Lock in the AFT.
E. Turn the channel selector to another station.
F. Release the AFT. Repeat steps B, C, and D.
G. Proceed as above for all channels.

III. Preset color level control (automatic control for color)

The preset color level control found on the front of color TV sets is designed to maintain color factors that are adjusted by a technician or controlled by a special TV transmission signal. Each TV manufacturer has its own name for this control. Under ordinary circumstances it should not be touched.

For emergency color adjustments:

A. Release the preset color level control.
B. Release the AFT.
C. Adjust the fine tuner to see if the color can be restored. Lock the AFT.
D. If the fine tuner fails to restore good color, adjust the color, tint, and brightness knobs to provide the most pleasing picture.
E. Watch the program.
F. Lock the preset color level control.
   Your adjustments do not affect the preset color level control. When it is locked again, the color will return to the original preset pattern.

GLOSSARY

AFT Automatic fine tuner.

Brightness knob Controls the amount of brilliance of the dark areas of the picture.

Color knob Controls intensity of colors, that is, a lot of color versus little or no color.

Contrast Sometimes called picture control. Controls the amount of brightness in the lighter areas of the picture.

Digital TV A new concept which refers to the way in which the incoming TV signal is processed within the set. With conventional analog TV sets, signal processing depends on variations in electric current. With the new concept, the incoming analog signal is converted and coded into binary numbers. The information can be stored in the set, called up any time and reconverted into a conventional analog signal for viewing.

Receiver Another term for television set.

Tint or hue Controls the color (blue, green, red) and allows you to get proper flesh tones.

25 Video

CONTENTS

Operating Tips
    General........................ 220
    Recording..................... 222
    Tape 224

Problems
    General........................ 225
    Recording..................... 227
    Picture 228
    Sound......................... 231
    Tape 232

InDepth
    More About Tapes.............. 233
    Cleaning Videotape Equipment ... 234
    Connecting Equipment........... 235
    Recording Procedures............ 235
    Playback Procedures 244
    Notes on Editing 247
    Erasing Tape 248
    Compatibility of Equipment and
        Tape Formats 248
Glossary......................... 249

Video systems have been commonly used since 1970. There are about thirty manufacturers of video systems, and over 120 models of the ½-inch video units. In addition to being used by broadcasting stations, video equipment is now used in education, business, and home entertainment, as well as in many other areas.

In order to use only one format, you may choose to transfer your slides, filmstrips, or super 8 and 16mm films to video. You can get the names of film transfer companies from your audio/visual dealer.

Caution: Videotapes, like other forms of communication, can be copyrighted. Unauthorized recording of such material may be contrary to the provisions of U.S. copyright laws.

A simple video system consists of a camera, a recorder, a monitor, and the necessary interconnecting cables. As of now, there are two main types of systems in the United States—helical-scan and quadruplex. The systems differ in the way the tape is scanned by the recorder video heads. The helical-scan system uses ¼, ½, ¾, and 1-inch tape widths. Most nonprofessional equipment uses ½ or ¾-inch tape. Quadruplex has been the professional system used in broadcast TV. It is large, expensive, and uses 2-inch tape. It is rapidly being phased out and replaced by the 1-inch helical scan.

Tape width affects the tape's storage capacity. A 1-inch tape stores more electronic information than a ½-inch tape. It produces clearer images, greater picture stability, and better details. On the other hand, the small ½-inch system is less bulky.

Video recorders come in different formats—open reel, cartridge, and cassette. The open-reel format requires you to thread the recorder, and the tape is not as well protected as in the other two formats. While used in the 1960s, few open-reel units with ½-inch tape are now used.

Cartridge units consist of one reel; the take-up reel is in the recorder. The cartridge has to be completely rewound before it can be removed from the machine. The cartridge format never became very popular and is not commonly used today. The information in this chapter relates only to open-reel and cassette formats.

A cassette consists of two reels enclosed in a plastic housing. It can be removed from the machine at any point in the program. As exists with the audio cassette format, the video cassette format comes in recorder-players as well as playback-only units. Tape widths for cassette systems range from ¼ to ¾ inch. The letter U or the term U-matic on a model number refers to a cassette format using ¾-inch tape (U-matic is a Sony trademark). The ¾-inch cassette format has been popular for business, educational, and professional needs. The ½-inch VHS and Beta cassette formats are particularly popular with the home market.

Manufacturers have been striving for smaller, more portable equipment. The latest result of this effort is the camcorder. It is an all-in-one camera and recorder, the ultimate challenge to the super 8 movie camera. There is a ½-inch Beta system that uses the standard ½-inch cassette, as well as a ½-inch VHS system that uses the compact VHS-C cassette, which can be placed in an adaptor and played on any standard VHS unit.

In an effort to provide an even smaller format than ½-inch, manufacturers are introducing the 8mm system. Eight millimeters refers to the width of the tape; the cassette housing is the size of an audio cassette. The 8mm format is available both in camcorders and in separate miniature camera and recorder systems. The cameras have as standard or have as options most of the features of the ½-inch units. Programs on 8mm or ½-inch formats can be transferred to the other format. There are also provisions for recording off-the-air. If you are thinking of the ultimate in portability, compare the various 8mm systems on the market, including available options and recording times.

Video equipment varies in weight. Traditional recorders range from 8½ to 100 pounds, while cameras weigh between 4 and 50 pounds. If a camera weighs more than 15 pounds it should be mounted on a monopod or a tripod, as it is too heavy to be steady when hand-held. While portable units are lighter than individual components, they can total between 40 and 60 pounds. However, without options some new camcorders weigh less than 5½ pounds.

Video equipment is delicate. It is subject to damage from jolts, changes in weather, tobacco smoke, ashes, humidity, dust, salt air, and magnetic fields.

Prices of video equipment vary according to the sophistication of the instrumentation. The following factors are important:

For recorders: portability
                 space-saver features such as front loading versus top loading
                 black and white versus color
                 editing capability
                 open-reel, cartridge, or cassette format

For monitors:
- screen size
- black and white versus color
- versatility of equipment:
  Can it receive a TV broadcast and/or closed-circuit signal?
  Can it transmit the signal to a recorder, if necessary?

For cameras:
- picture quality
- type of electronic video pickup—tubes or the latest solid-state chip
- portability
- operational features and accessories
- special features such as various slow-motion playback speeds, timer for unattended recordings
- lens interchangeability with other cameras
- maintenance costs

In making your selection, consider compatibility and interchangeability of equipment and tape formats (see InDepth VIII).

Note that home TV units instead of monitors or monitor TVs can be connected to your VCR. Depending on recorder capability, you may watch VCR playback on your home TV, and you may watch a program on your TV while recording another. You may not record a TV program directly from your television set onto the recorder.

Tape prices vary according to the format, brand, number of minutes and quantity ordered. Determine your needs and compare prices before buying. Open-reel tapes come only in 30- and 60-minute lengths; ¾-inch cassette tapes come in 5-, 10-, 15-, 20-, 30-, and 60-minute lengths; ½-inch cassettes come in 30-, 60-, 90-, 120-, and 160-minute lengths. Other lengths for each format are available.

OPERATING TIPS

I. **General tips**

A. Camera:

1. After turning the camera on, it must warm up for at least ten seconds. Then remove the lens cap.

2. The part of the camera that changes incoming light into electric current is the pickup tube. Never point the camera directly at the sun or other source of bright light, as this can damage the camera's tube in many units, and leave a permanent image in the camera.

3. Avoid continuous shooting of a subject in strong light, especially when the picture has high contrast.

4. The smaller the lens opening (higher the f number), the more light is needed to get a good picture. On many cameras the lens opening can be set either automatically or manually,

5. When the camera is not in use, keep the f stop "closed" and the lens cap on so there is no chance that a bright light can damage the pickup tube.

6. Color cameras have a WHITE BALANCE switch or button which will adjust camera color to the kind of light available. You may then fine tune the color to your liking, using the color adjustment dials where provided.

7. Where switch is available, set

INDOOR/OUTDOOR or CLOUDY/SUN for appropriate conditions.
8. The brightness and contrast controls on the viewfinder usually only affect the viewfinder, not the video output signal from the camera.

B. Recorder:
1. Never operate the machine right after having transported it from a cold location directly to a warm location. Wait about one hour before using it. Also check the DEW indicator light, if available.
2. Traditional portable units usually can be operated either vertically or horizontally. For example, on a field trip you may be able to carry the recorder on your back.
3. In video cassette recorders, do not turn off the power switch while the tape is threaded in the machine. Stop the tape first.
4. For open-reel recorders, do not press down or place anything on the head drum cover. Do not touch the heads except to clean them, and do not touch the heads while the motor is running.
5. A recorder must have its own TV tuner to record off the air if it is not depending on input from the monitor TV set. It must have an RF modulator to play back a taped program through a regular TV set. Most units in the home video systems (Beta and VHS formats) have both of these features.
6. For most machines that record at various speeds, the machine will automatically set the proper playback speed.

C. Monitor, monitor TV, or TV:
Never remove the back cover of the set or perform internal repairs. You may be subject to X-ray radiation, electric shock, or fire hazard.

D. Other:
Should any material, liquid or solid, fall into the cabinet of the recorder or monitor, unplug the unit and have it checked by a qualified person. Otherwise, electric shock or fire may result.

E. Antenna:
For most recorders, in order to watch a program on your home TV set and record another, the antenna must be attached to the recorder, not to the TV. Then connect the recorder to the TV. In this arrangement, the antenna serves both the VCR and the TV. The arrangement assumes the VCR has its own tuner.

F. Cable:
1. If you are considering hooking your TV up to cable, check out how cable might affect the functioning of your VCR. You may need to rent an additional converter box if you want to watch one TV program and record another on your VCR. Your flexibility to program various channels in timer mode may be affected.
2. Before buying a VCR, see if a cable-ready VCR unit would be

useful with your community's cable system. It may or may not be.

II. Recording tips

A. Methods of recording:

Live presentations can be recorded using a camera and a VCR unit. You may also copy a friend's legally recorded tape. Off-the-air recordings of TV programs are also possible: If necessary, TV programs can be recorded from a monitor TV set, not a home TV. They can also be recorded from a VCR unit which contains its own tuner. When this type of VCR is connected to your TV set you can record one program on the VCR and watch another program on the TV.

B. Before starting to record:

1. Check out the system. Record a small segment of audio and video, and play it back to check focus and other video qualities.

2. Make sure you have proper accessories:
   - spare connecting cables
   - spare plastic take-up reel for open-reel recording
   - 3-2 adaptors for AC outlets in old buildings
   - extension cables

3. For recording TV programs off the air, where possible check to see that you have a clear, stable picture.

4. For portable units, make sure all batteries are fully and freshly charged. It takes several hours to completely charge a battery. For ½-inch cassette formats, a fast charger can recharge a battery in two hours. Operating instructions are with the kit. Do not charge longer than the specified time. Check the battery after a couple of hours to see if it is holding the charge. If not, replace the battery through a store that carries videotape equipment. The cost range is approximately $30 to $75.

5. Check out your tape stock. You need a sufficient amount in good condition. If you are shooting with used tape, have it erased (degaussed) in a bulk eraser for best results.

6. Set recorder speed to determine the recording time for the length of tape you are using. Many recorders have the following tape speed positions: SP (standard play), LP (long play), and SLP (super long play). Like an audio reel-to-reel recorder, remember that the faster the speed, the better the fidelity.

7. Monitor your picture before a live recording to see that you have sufficient light. Lighting is necessary for good camera pickup, good picture composition by the photographer, and establishing the proper mood. Good lighting will also reduce picture graininess and increase details. Consider using a keylight, a backlight, and a fill light. See Glossary.

When buying lights for color cameras, note that all video cameras are manufactured for 3200-degree Kelvin (3200°K) lights.

Black-and-white recordings can be made in lower light levels than color. Normal classroom-type light is sufficient for most black-and-white and recent color video cameras. Older color cameras probably need supplementary lighting.

C. Making adjustments to record.

Adjustments should be made with the equipment on and fully warmed up. On many units, the picture will appear on the camera viewfinder approximately ten seconds after the power is turned on.

1. If recording off the air from a monitor TV set into a recorder:
   Make the adjustments in the monitor TV set. In most cases, if you are receiving a satisfactory picture after adjusting the monitor TV set, you do not need to touch the recorder controls.
2. If recording off the air using a cassette recorder with its own tuner:
   Make adjustments primarily in the recorder unit. A tuner works like a TV set, without the picture tube. Most home cassette units have tuners.
3. If recording live:
   You may be able to adjust the sound on the recorder, although now recorders usually do not provide manual audio adjustment because the audio level is controlled automatically. Before recording, turn the monitor volume up to make sure you are getting proper sound quality. Turn the volume down before starting recording to eliminate feedback. With a portable recorder, use earphones to monitor sound because you cannot hook up the recorder to a monitor while recording.

With most cameras, adjust the video by pointing the camera at a scene or card with 50 percent white area. Then make adjustments in focus and lens opening (f stop), where possible. Most cameras can operate over a wide range of lighting conditions.

Adjust the lens opening (f stop) as follows:

- Indoor shots with normal artificial lighting: f 1.8–f 4
- Outdoor shots with clouds or shade: f 8
- Outdoor shots having a bright scene: f 8–f 16

On most cameras, to focus a zoom lens at a particular distance,

- Adjust the f stop for appropriate lighting conditions.
- Turn the zoom ring all the way to closeup.
- Focus.
- Now use the zoom.

D. While recording:

1. General information. Do not record at the very beginning or end of a tape, especially on open-reel recorders. The ends are where tapes most often get damaged. After turning on the equipment, use up five to ten seconds of tape before recording your program. Before recording at the beginning of a tape, set the tape coun-

ter to "000". Also, identify the particular segment by an audio or video slate, indicating such pertinent information as subject and date. This is particularly important if you plan to have several segments on one tape.

Keep a log or written list of scenes that have been shot. Use the counter numbers from the recorder to indicate the beginning and end of a scene.

To change the mode of operation on many recorders, press the STOP button, wait for the STANDBY light to go off, where there is such a light, and then change the mode.

Allow five to ten seconds after taping action before pressing STOP.

Whenever possible during taping, play back your tape or the first and last fifteen seconds of each take to make sure that all is well. Then fast forward a few inches to avoid recording over your taped material.

2. Open-reel recording tips. When you are ready to tape, let the recorder roll for five seconds before starting to use audio or video slate. (see Glossary)

When you're finished recording or playing back a tape, rewind it completely and remove it from the recorder.

3. Cassette format tips. If you are using the same audio channel (MIC INPUT) with both a microphone and sound being fed in by another electronic input such as monitor or tape recorder (LINE INPUT), the live microphone has priority, and the other input will not be recorded on that same audio channel. Thus, in order to record both effectively, you would need to use two audio channels where available.

When the tape reaches its end, it will stop automatically. In some recorders, it will be rewound automatically to the beginning, will fast forward for a few seconds to wind up leader at the beginning, and will then stop.

Remove the cassette before turning the power off, since the cassette compartment in many units needs to rise with the power EJECT button.

The cassette tape counter can be erratic. For replay at a specific spot, check the picture visually. On some units, use the counter only as a general indicator for location of a segment because the counter may not be very precise. Automatic tape stop may occur at any time during recording or playback when improper tape motion occurs (tape-protection device).

III. Tape tips

  A. General information:

   1. Do not force the cassette tape into the recorder unit. It will only fit in one way.
   2. The same videotape will record both black and white and color—the camera makes the difference.
   3. The shiny side is the emulsion or recording side.

  B. Cassette tapes. Safety feature to prevent accidental erasure.

For ¾-inch tapes, there is a small safety button on the bottom of the cassette. If this is removed, you cannot record or perform audio dubbing. To reuse a tape for recording, reinsert the safety button. As you record, previous recordings are automatically erased. This button, whether in or out, has no effect on playback.

On ½-inch tapes, there is a tab on the video cassette that functions like the safety button for ¾-inch tapes. Break the tab to prevent erasure of a program. Cover the punched-out area with masking tape to record again.

C. Open-reel tapes:
1. The hub or center is the strongest part of the reel. Always handle by the hub and not the edges (also called flanges). Damaged edges or distorted reel will damage the tape.
2. With clean, dry hands, thread the recorder. Check the threading diagram before starting.
3. Never thread the recorder while the heads are spinning. Wait for them to stop.
4. Keep an eye on both reels to make sure they are moving smoothly. If the tape gets clogged, undo carefully, cutting tape only as a last resort.
5. Wrinkling or touching the tape itself too much will damage the tape.

D. Proper tape identification:
Label the reel or cassette and box. Indicate date recorded, title, performers, and person who recorded it.

E. Storage conditions for tape:
1. Store open reels and cassettes under moderate temperature and humidity conditions. A cool, dry, dust-free place would be ideal.
2. Store tapes upright; do not lay them flat.

PROBLEMS

PROBLEM	SOURCE	SOLUTION
General Problems		
1. Black screen, no picture, no light, all power gone.	Monitor or TV set.	Check AC power cord of monitor or TV.
	AC adaptor cord with portable recorder.	AC adaptor may be plugged into outlet, but switch still on BATTERY. Move switch to AC. Be sure AC adaptor is turned on.
	Battery for portable recorder.	Recharge battery or use AC power.

226   *Video*

PROBLEM	SOURCE	SOLUTION
General Problems		
	Recorder.	Check connection to AC outlet.
		Check DEW indicator, where available.
		Machine should be ON. PAUSE OFF.
		Check tape threading.
		If TIMER is on for later recording, cannot operate manually. Switch to POWER.
		Blown fuse. Take in for repair.
2. Blank white screen, lit up, no audio, no video.	Monitor or TV.	Monitor or TV set not hooked up to recorder.
3. No off-air recording made. See also Problem 10 a.	½-inch cassette recorder format.	Remove cable in VIDEO INPUT and AUDIO INPUT jacks used for recording with camera.
	Antenna.	Check that antenna is properly connected.
	Timer in recorder.	Retrace steps for setting timer.
		Power switch: With normal recording power ON, timer OFF. With timer recording power OFF, timer ON.
		On many units, a one-second interruption before or during recording will stop timer recording.
	Tape.	Tape missing tab or safety button.
4. Screen blank, except for very bright line.	Monitor or TV set.	Turn off immediately. Take in for repair.

5. Frequent snapping or popping sound when set is already warmed up.	Monitor or TV set.	Unplug set. Take in for repair.
6. Battery does not hold a charge.	Portable recorder.	Discharge battery all the way down. Recharge battery completely. If battery still does not hold charge, replace.
7. Recorder will rewind slowly but will not go forward.	Portable recorder.	AC adaptor not connected to recorder and AC outlet.

Recording Problems

8. Picture cannot be focused in camera during live taping.	Battery for portable recorder.	Battery run down. Recharge battery.
9. Diagonal lines in camera viewfinder during live recording.	Battery for portable recorder.	Recharge battery.
10. No picture in camera viewfinder.	Camera.	Check camera cable for breaks and proper connections.
		Turn camera ON.
	Recorder.	Turn recorder ON.
		Switch TV setting to CAMERA, where appropriate. Check cable connections.
	Battery for portable recorder.	Recharge battery or use AC power.
10a. No recording being made (see also Problem 3).	Recorder.	Check that power is ON, unless using timer.
		Release PAUSE button, if on.
		Press both RECORD and PLAY.
		For ¾-inch cassettes, safety button must be inserted to depress RECORD and PLAY. For ½-inch cassettes, punched-out safety tab must be covered with tape to depress RECORD and PLAY.

228   *Video*

PROBLEM	SOURCE	SOLUTION
General Problems		
		Make sure there is a cassette in VCR.
		Check DEW indicator.
		Cassette tape side full. Flip cassette to other side.
		Relock cassette tape compartment in recorder.
	Off-the-air recording from monitor TV, using VCR without tuner.	Turn monitor TV ON. Also check InDepth IV D.
	Recording from camera.	Check power supply connections (see InDepth IV A).
Picture Problems		
11. White flashes in picture.	Recorder or monitor TV during recording.	Interference during recording from fluorescent light, large motors, faulty appliances. Find source for future recordings. Recording problem cannot be eliminated on playback.
	Monitor TV or TV on playback.	Locate source of interference.
12. Snow concentration with or without picture.	Recorder.	Clean heads and tape path. Cleaning should be done by a qualified person.
		On playback, slowly try to rotate TRACKING control on recorder.
		Weak signal when recorded; cannot cure.
		Switch VTR/TV to VTR, or VCR/TV to VCR.

Video **229**

	Monitor TV or TV set.	Check antenna connection. Locate local interference.
		Check that your recorder and monitor TV (or TV) are properly tuned.
	Antenna.	Weak signal for recording. May need more powerful antenna.
	Cables.	Retrace all cable connections.
	TV set.	For playback, TV channel should match channel on recorder RF adaptor (see InDepth V A).
	Cassette tape.	Tape missing tab or safety button; no recording made. Record again.
13. Snow or breakup at certain points, unstable picture.	Recorder.	Have dirty heads professionally cleaned.
	Tape.	Buy better-quality tape.
14. Cannot get just-recorded picture on playback. Previously recorded image still appears.	Recorder.	Check recording procedures. Record again as follows. Leave safety button in the cassette. In most cases, hold RECORD button down when pushing PLAY button.
15. Poor contrast; washed-out picture.	Monitor TV set, or TV.	Check automatic fine tuning, if TV has control. Check brightness and/or contrast control. Check color control.
16. Bars tilted left or right. Picture tears sideways.	Monitor TV set, or TV.	Adjust HORIZONTAL HOLD control, where available. Check for local interference from appliances. If unable to correct, take in for repair.
	Recorder.	Slowly adjust TRACKING control.

PROBLEM	SOURCE	SOLUTION
Picture Problems		
17. Vertical roll, upward or downward.	Monitor TV set, or TV.	Adjust VERTICAL HOLD control, where available.
18. Top of picture bends over.	Recorder.	On playback, turn SKEW control, where available. When recording, return knob to center position.
	Monitor TV set, or TV.	Adjust HORIZONTAL HOLD control, where available.
19. Distortion or interference.	Recorder.	Clean heads.
		Put COLOR/B&W switch in appropriate position, where applicable.
		Slowly adjust TRACKING control on recorder.
	Tape.	Rethread tape on open-reel recorder.
		Tape old, damaged, or defective. Replace and record again.
20. Distortion; partial erasure of tape.	Recorder.	Have heads professionally demagnetized. Cannot improve already recorded tape.
21. Picture on playback shows some just-recorded portions and some old programming.	Recorder.	Recorded with used tape. If recording live and FAST FORWARD is presssed on used tape, that section keeps old programming. To correct, record over these sections.
	Cassette recorder.	Allow ten to twelve seconds of tape lead time before recording program and at end of recorded segment.
22. Ghosts (double image).	Monitor TV set, or TV.	Adjust fine tuning.

	Antenna, during recording.	Reposition antenna.
		Experiment with different antenna orientations.
		Install directional antenna, accepting TV signals from one direction only. If using cable, call cable company.
	Cable company.	
	Cables to connect equipment.	Use another set of cables.
23. Picture out of focus on screen or permanent black spots in picture after recording with camera.	Monitor TV set, or TV.	For out-of-focus picture, adjust fine tuning.
	Tube of camera.	Clean or replace camera tube.
23a. Cannot watch normal broadcast on your TV, when TV is connected to recorder.	Recorder.	Switch VCR/TV to TV. Also check antenna connections (see InDepth IV B).

Sound Problems

24. Sound, no picture.	Monitor TV set, or TV.	Check fine-tuning control. Check brightness control.
	Recorder.	Check cable connections and input switches on recorder.
25. Picture, no sound.	Recorder.	If using remote microphone for live recording, replace auxiliary microphone cable, check connections.
		If recording off-the-air, disconnect microphone.
	Cassette recorder on playback.	Where available, try audio channels on CHANNEL 1, CHANNEL 2, or MIXED.
26. Sound with extraneous buzz or hum; broken-up picture.	Monitor TV set or TV.	Local interference. Try to turn off nearby appliances.

232   *Video*

PROBLEM	SOURCE	SOLUTION
Sound Problems		
	Tape.	Hum on original tape cannot be corrected in playback.
	Cords and cables.	Check power cord.
		Check cable connections to recorder.
27. Distorted sound.	Monitor TV set or TV.	Check cable connections to recorder. Also see if cable is broken.
	Microphone.	Check microphone cable and microphone battery where there is one.
		Keep microphone away from monitor TV set or TV during live recording.
	Recorder.	Audio controls should be turned up or set on automatic level control. Record again.
	Tape.	Try prerecorded tape that you know has good sound and picture.
Tape Problems		
28. Tape gets stuck.	Cassette recorder.	Leave tape alone or remove carefully; take recorder in for repair.
29. Tape breaks.	Portable open-reel recorder.	Do not splice. Throw away shortest end; rethread.
	Half-inch open-reel recorder format	Do not splice. Throw away shortest end; check threading path and rethread. Take recorder in for repair.

30.	Tape unthreads and stops; STANDBY lamp lights briefly and goes out.	Cassette recorder.	On many recorders, try the following: Press REWIND to return to beginning. Press FAST FORWARD to near end. Try to play back for one minute. If operation restored, continue. If not, try another cassette. If tape still stops, take machine in for repair.
31.	Tape stops and remains threaded. STANDBY lamp stays on, where available. Machine does not operate in any mode.	Cassette recorder.	Turn off machine. Let it stand for a half-hour. If not restored, take it in for repair.
32.	Tape stops, recorder stops.	Open-reel recorder.	Check threading path.

IN-DEPTH

I. More about tapes

A. Tape Damage:

Tapes have the ability to retain information for an indefinite time. The signal on the videotape begins to deteriorate after repeated takes. The tape can be contaminated by food, drink, or ashes, altered by an external magnetic field, and damaged by heat. Tapes and their cassettes can be damaged by being put on the dashboard or in the glove compartment or trunk of a car.

Tape does not burn easily. However, there may be some distortion or darkening of the tape. If it has been exposed to fire, the tape should be rewound at medium tension, then the signal transferred to a good tape.

B. Shipping Tapes:

The container should be rigid and strong to protect tapes from dropping or crushing. It must have some water resistance, in case it is left out in the rain.

Tapes should be packed with bulk spacing material, such as wood or cardboard, between the tape boxes and the outer shipping container. Tapes in transit may be subject to temperature extremes. They should therefore be allowed to reach room temperature before being used—approximately one hour.

When shipping tapes by air, mark them "Caution—Do Not X-Ray." Otherwise, the tapes will be ruined.

C. Buying Videotapes:

1. Use good-quality tape; otherwise, the quality of your pic-

ture will suffer. Name brands are usually a safe bet.

2. High-density tape and low-noise tape provide better picture quality. Almost all tape today is of this type.

3. Extended-play tape provides extended recording and playback time. This tape is thinner than average, may not last as long, and may break. As with audiotapes, extended-play tapes are not recommended when they will be used repeatedly or mass-duplicated. Also, they may not operate properly in the machine.

II. **Cleaning videotape equipment**

A. Cleaning Video Recorders:

1. General information. Heads are the devices on recorders that read, scan, or look at the information to be recorded or played back. Occasionally, the heads get dirty with debris from the tape. Cleaning the heads on a cassette recorder is not difficult. However, cleaning the heads on an open-reel recorder is a delicate operation. Demagnetizing the heads is not necessary often. However, when the picture is erased, this could be the problem. Consult your dealer.

2. To clean the cassette heads, use video cassette-cleaning tape. Insert it like an ordinary video cassette. Reset the tape counter to 000. Push PLAY and watch the tape counter. Press STOP at five seconds. Remove the cassette. Do not rewind the cassette cleaner at the end of each use. However, it may be totally rewound and reused several times. Do not use the cleaner unless picture symptoms indicate the need for video head cleaning.

Excessive use will shorten head life substantially.

There are two types of cleaning cassettes: a cassette that is inserted into the recorder and cleans as above, and a cassette inserted into the recorder and used in conjunction with a monitor or TV set. A color pattern on the monitor (similar to that used by TV stations to sign on and off the air) shows the effect on the color as the heads are being cleaned. If the color does not improve, you know the problem is not in the heads of the recorder.

When purchasing a cleaning cassette, give your dealer the make and type of your equipment, so that he or she can recommend the proper cassette cleaner. Some are more abrasive than others.

3. Cleaning open-reel heads. There should be routine professional cleaning of open-reel heads and tape path.

B. Cleaning Monitor or TV Set:

1. *Picture tube.* The set must be turned OFF when washing the picture tube. Use mild soap and water-dampened, but not wet, cloth. You can also use a glass cleaner. However, do not use aerosol directly on the tube face. Rinse and dry thoroughly.

2. *Cabinet.* Clean with a polishing cloth. Never use solvents such as thinners or acetone.

C. Cleaning Camera:

Turn the camera OFF before cleaning.

1. Clean the lens with camera lens tissue or liquid lens cleaner. Never spray liquid directly onto the lens. Place a drop of liquid on the lens tissue.
2. Use a soft cloth to clean the camera body. Do not use liquids.
3. Clean the camera pickup tube, where possible, particularly when black spots appear in the picture, as follows:

    • Detach the lens from the camera.

    • Use lens cleaner, moisten a cotton swab, and gently wipe the tube.

    • Blow compressed air on the tube to remove specks of dust. Test compressed air on your hand first to remove any moisture in the can.

    • Check with dealer if black spots remain in the picture after cleaning the tube and the rest of the camera.

III. Connecting equipment

Plug types are not standardized for each type of connection. For example, with two different brands of recorder, you cannot assume that a certain type of input or output will always use the same kind of plug. Check your recorder socket connections before buying cables.

Do not assume that multi-pin plugs are all interchangeable. Equipment connections must match physically and electronically. The rule of thumb is: Don't make assumptions when connecting video equipment. If in doubt, consult your dealer.

When using separate audio and video cables, they should be about the same length.

Shown below are some commonly used connectors (Figure 25–1).

IV. Recording procedures

Below is a brief outline of ways to record. Make your choice, then turn to the appropriate section. Because there are dozens of different models on the market, each with its own idiosyncrasies, use the instructions in this section to understand the basic procedures for recording and playback. Then apply these concepts to your own equipment.

All switches must be off before making any connections. Be sure the recorder TRACKING control is at the center position. Allow yourself about twenty minutes to set up and check out your equipment before actually taping (see Figure 25–2).

Methods of Recording:

- Live recording with camera and recorder. (See A, page 238.)
- Off-the-air recording—watching one program on your TV set while recording another on the VCR. Applicable to most home unit situations. (See B, page 241.)
- Off-the-air recording when the recorder contains its own tuner. Applicable to most home units and to timer recording. (See C, page 242.)

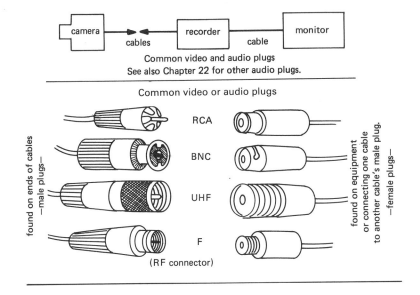

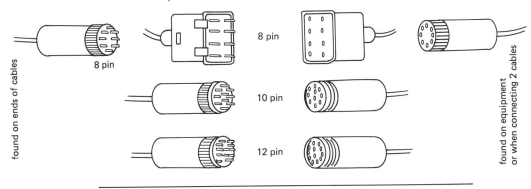

Figure 25-1 *Basic plugs for connecting equipment.*

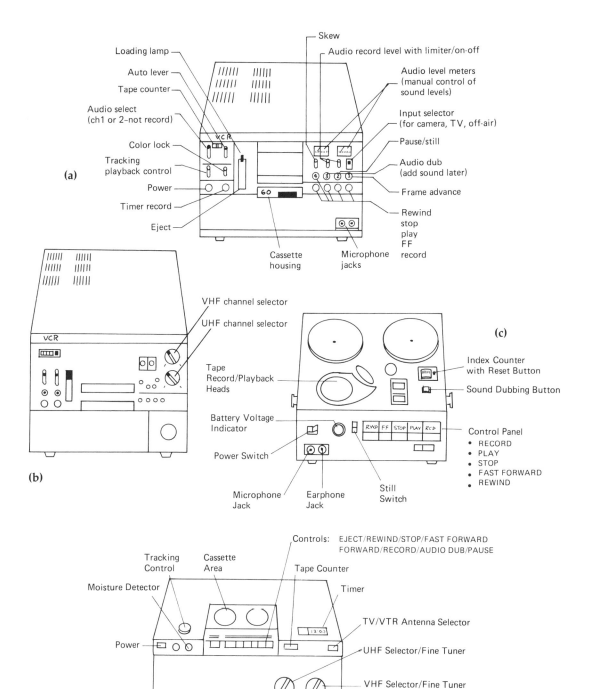

Figure 25-2 *Recorder/player units. (a) 3/4" cassette. (b) 3/4" cassette with built-in tuner. (c) 1/2" portable reel-to-reel. (d) 1/2" cassette unit.*
*Most units come with fine-tuning controls for each channel. Many units have timer feature for unattended recording.

- Off-the-air recording from monitor-TV unit to recorder which does not contain its own tuner. Not for use with a regular TV set. (See D, page 243.)
- Copying tapes. (See E, page 244.)

A. Live recording with camera and recorder:

(See Figure 25-3.)

For monitoring the program, include step 2.

Connections:

1. Connect the recorder to the AC outlet.
2. Monitoring the program (Battery-operated equipment will not monitor the program while recording is taking place.) Connect the monitor or monitor TV to the AC outlet. Connect the monitor TV to the recorder. For connections appropriate for your equipment, see Figure 25-7, page 246. Turn equipment ON. Where available, set TV/VCR to VCR on monitor TV (to monitor a live recording on your home TV, see InDepth IV B Steps, 1, 7).
3. Check the camera power supply. Depending on the manufacturer, make connections as follows (see Figure 25-4):
   - The camera may have its own power supply and plug into a wall outlet.
   - The camera may not have its own power supply and plug into the recorder only.

Figure 25-3 *Basic parts of a video camera.*

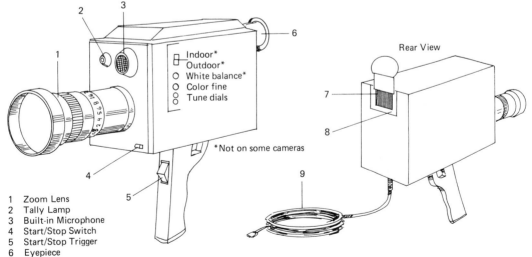

1 Zoom Lens
2 Tally Lamp
3 Built-in Microphone
4 Start/Stop Switch
5 Start/Stop Trigger
6 Eyepiece
7 Viewfinder
8 Recording Lamp w/Battery Power Warning through Viewfinder
9 Camera Cable

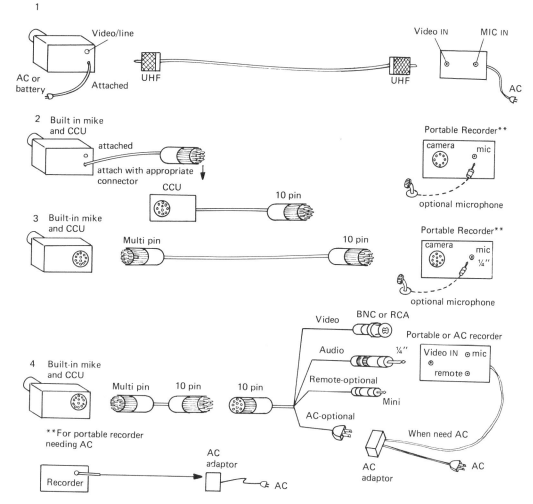

Figure 25-4 *Four different camera/recorder connections.*

4. Connect the camera cable to the recorder terminals as follows: Use either CAMERA INPUT or CAMERA of recorder, or both VIDEO INPUT and AUDIO INPUT of recorder. If both sets of jacks are available in the recorder, decide according to whether the camera cable has two plugs or one plug at the end.

5. If using a separate microphone, usually connect the microphone cable to MIC INPUT of the recorder. If using only one microphone on a two-channel system, use channel 2 of recorder. Using a remote mike will disconnect a built-in microphone.

Video 239

Recorder:

6. Turn the recorder ON. Where available, be sure TIMER switch if OFF.
   Set the VTR/TV (VCR/TV) selector to VTR (VCR). Set the tape counter to "000." (Where available, set CAMERA or LINE/TV input switch to CAMERA or LINE.)
   If using ½-inch VHS or Beta formats, set the recorder tape-speed selector for the amount of taping time desired.

7. Insert the tape into the recorder.
   For reel-to-reel machines, thread the tape according to the diagram on the recorder. Tape should unwind counterclockwise with the shiny side toward the heads.
   For cassette format, make sure a recording can be made.

   - On ¾-inch cassettes, the colored button must be in the cassette hole provided; otherwise, obtain one and insert.
   - On ½-inch cassettes, the tabs must not be punched out; otherwise, cover the holes with masking tape.

   Press EJECT and insert the tape.

8. Audio recording levels are usually automatically set and maintained in the recorder. In some machines, it is possible to set the sound level yourself.

Camera:

9. Set the camera switch to STANDBY.
   When possible, set the lens f stop to the automatic setting or automatic iris.
   For the manual setting, set the f stop to f16. Remove the lens cap. Look through the camera viewfinder, focus the lens, and adjust the f stop (open or close the opening of the lens) for the particular lighting conditions.

Test Equipment:

10. Record a short segment and play it back to make sure you are recording properly. On playback, where possible, look at the picture through the viewfinder or use playback procedures.
    To test equipment:

    - Press RECORD and PLAY on the recorder.
    - Squeeze the trigger or turn the camera switch, if available; otherwise, activating the recorder will activate the camera.
    - After pressing RECORD and PLAY, wait five seconds before recording so that the tape is rolling properly. With ¾-inch cassette format, some machines lower the cassette automatically. Wait about fifteen seconds before recording.
    - Record segment. Press STOP. Play back segment. Check quality.

To record:

11. If there is no problem with recording, rewind the tape and reset the tape counter to "000." Then begin recording, as explained in step 10 above. Make a slate at the beginning of your tape to identify this program (see Glossary).

To end recording, and to play back:

12. To end the recording, press STOP. Rewind the tape. Some ½ inch and ¾ inch cassette format tapes automatically rewind the tape to start when the tape reaches the end.

13. For playback procedures, see InDepth V.

B. Off-the-air recording—watching one program while recording another:
This is applicable to most home units. The VCR has its own tuner.

Connections:

1. Before buying the VCR, the antenna went from the roof or rabbit ears directly to the TV. Now the antenna needs to go from the roof, or separate rabbit ears, to the VCR and then to the TV so the antenna can serve both units.

   Disconnect the antenna from your TV. Attach it to the VCR antenna terminals (for example, VHF IN, UHF IN). (See Figure 25–5.) Run the appropriate cable from the VCR antenna terminals (VHF OUT, UHF OUT) to the TV antenna terminals.

2. Connect the VCR and TV set to AC outlets.

Recorder:

3. On the recorder, dial the UHF or VHF channel from which you want to record. Be sure TIMER is OFF, where applicable.

4. Set the selector switch on the recorder to TV, not VCR. Set tape speed, where feature is available.

5. Follow steps 4 through 6 in section C.

Figure 25-5 *One type of off-the-air recording connection for recorder with built-in tuner.*

TV:

6. To watch another program while recording, turn the TV on and select the station you wish to watch. Watch it! (If you are not recording with the VCR, but just wish to watch a TV program on your television set, keep the recorder switch on TV, and select the station on your television as above.)
7. To monitor the recording rather than watch another TV program, set the switch on your recorder to VCR instead of TV. On your TV use channel 3 or 4, 5, or 6, whichever is marked and set on the back of the recorder and is not in use in your area. Leave the VCR unit on the channel you selected for recording the program.

C. Off-the-air recording when the cassette recorder contains its own tuner: No monitor TV is essential for recording.
Applicable to most home units and to timer recording.

- *Regular, non-timer recording.*

Connections:

1. If not already connected, attach the antenna to the tuner on the back of the recorder. One way is shown in Figure 25–5. The connections you select depends on the terminals on your equipment. Connect recorder to AC outlet.

Recorder:

2. On the recorder, dial the VHF or UHF channel from which you want to record. Where available, be sure TIMER is OFF, unless using timer recording below.
3. Set tape speed, where feature is available. Set the switch on the recorder to VCR TUNER or VCR.
4. To make sure a recording can be made, check your cassette tape. On ¾-inch cassette format, the colored safety button must be in the tape; if it is not, obtain one and insert. On ½-inch cassette format, the tabs must not be punched out; if they are, cover the holes with masking tape.
Press EJECT. Insert cassette.

To record:

5. Press RECORD and PLAY at the same time. Wait five seconds before recording so that the tape is rolling properly. With ¾-inch cassette format, some machines lower the cassette automatically. Wait about fifteen seconds before recording.

To end recording:

6. Press STOP. Rewind the tape. For playback procedures, see InDepth V.

- *Timer recording:* (If VCR is not already connected to antenna, see above)

1. Set TIMER ON, power OFF. Power will be turned on and off by timer.
2. Set the correct actual time on your unit's clock.

3. Set the day to make the recording and time to start recording.
4. Follow steps 2, 3, and 4 for regular recording above. Do not use RECORD and PLAY at all. This is controlled by the timer mechanism.

   Note: When the timer is ON, you cannot operate the unit manually. To stop automatic recording, release TIMER switch.
5. Set time for recorder to stop recording.

D. Off-the-air recording from monitor TV unit to recorder without its own tuner.

Not for use with a regular TV set.

Connections:

1. Connect the monitor TV to the AC outlet and to the antenna. Connect the recorder to the AC outlet.

   For a battery-operated recorder being used on an AC outlet, connect the AC adaptor to the recorder and to the AC outlet. Flip the AC adaptor switch to POWER to disconnect battery power.

2. Connect the monitor TV and the recorder to each other. Use one of the ways in Figure 25–6, depending on the equipment and the connecting cable.

Figure 25-6 *Off-the-air recording connections—monitor/recorder.*

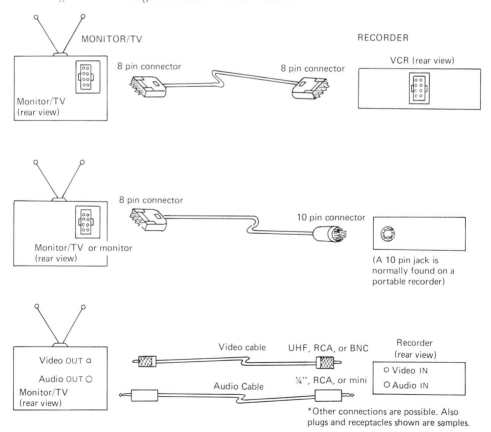

Recorder:

3. Set the switch to TV, not CAMERA or LINE. If features are available, set tape speed and be sure TIMER is OFF.
4. Insert the tape into the recorder.
   For reel-to-reel machines, thread the tape according to the diagram on the recorder. Make sure a recording can be made: On ¾-inch cassette format, the colored safety button must be in the tape; if it is not, obtain one and insert. On ½-inch cassette format, the tabs must not be punched out; if they are, cover the holes with masking tape.
   Press EJECT. Insert the tape.

Monitor TV:

5. Turn the set ON. Set monitor TV to VCR.
6. Select the channel from which you want to record.

To monitor program:

7. Monitoring the program will indicate that the recorder is receiving the picture. However, only playing back the tape after recording a portion of a television program will show whether a recording was actually made on the tape.
   On some machines, press RECORD on recorder. On some newer machines, press STOP on recorder for the picture to appear on the monitor TV.

To record:

8. On most machines, press both RECORD and PLAY at the same time.

To end recording and to play back:

9. Press STOP on the recorder. Use playback procedures below.

E. Copying:

Do you want to copy a friend's cassette?

Use two VCR units, the first (VCR 1) to play back the original, and the second (VCR 2) to record the information onto a new tape. Obtain the proper connecting cables by telling your dealer the make and model of each unit. For best results, connect:

Playback Machine (VCR 1)	VIDEO OUT/ AUDIO OUT
to	to
Recording machine (VCR 2)	VIDEO IN/ AUDIO IN

Insert the cassettes in the machines. Press RECORD and PLAY on the recording machine, then press PLAY on the playback machine. It would be helpful to hook up a monitor or TV to the recording machine to check what is being recorded.

V. **Playback procedures**

It is important for all switches (including TIMER switch, where applicable) to be OFF before making any connections.

Set recorder VCR/TV selector to VCR.

A. *Playback with the average home TV set:*

If the video recorder-player has a special feature (RF adaptor) you may use a regular TV for playback of taped programs. Almost all home recorder-players are so equipped.

1. If not already connected, connect the TV and the recorder to the AC outlet. Then connect the TV and the recorder to each other (see Figure 25–7, methods 4 and 5).
2. Select the appropriate playback channel as follows:

    On your recorder: Check the back or side panel of your recorder. Look for Channel 3 or 4, 5, or 6, and select the channel not in use in your viewing area (you may leave this channel setting on all the time). Turn unit ON.

    On your television set: Dial the VHF selector of your TV to the playback channel just selected on your recorder. After making a channel selection as explained above, turn the TV ON.
3. Use the VCR setting on the recorder-player, where there is an input switch.
4. Insert the cassette or thread the tape. Press PLAY or set the machine in FORWARD.
5. Adjust the TV for the best picture.
6. After playback, equipment should usually be turned off, disconnected and unplugged from the AC outlets. Tapes should be stored vertically.

B. *Playback with monitor or monitor TV:*

In places where there is a central antenna system, you may find it necessary to disconnect the monitor TV set from the antenna in order to get good reception on your video playback. Sometimes the outdoor signal will overpower the tape signal.

1. Connect the monitor or monitor TV and the recorder to the AC outlet. In some cases you have only a player not a recorder-player; however, the connections are the same.
2. Connect the monitor or monitor TV and the recorder to each other (see Figure 25–7, methods 1, 2, and 3 for a monitor and methods 1 to 5 for monitor/TV).

    With a monitor TV set the input selector switch to VCR or EXT (external). Look for the switch on the front, back, or sides of the unit. Turn the monitor TV ON.

    (If using a TV-type connection for your monitor TV, see InDepth V A 2.)
3. Insert the cassette or thread the tape.
4. Press PLAY or FORWARD on the recorder.
5. Adjust the monitor or monitor TV for the best picture.
6. After playback, unplug equipment and store tapes vertically.

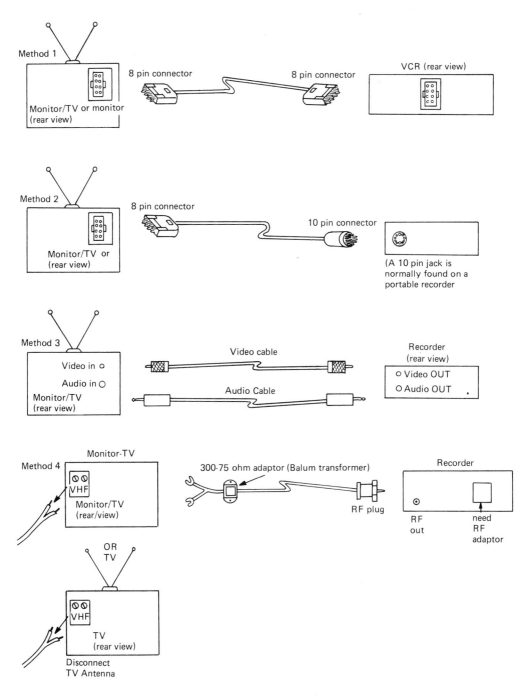

Figure 25-7 *Playback or monitoring connections—monitor/recorder.*

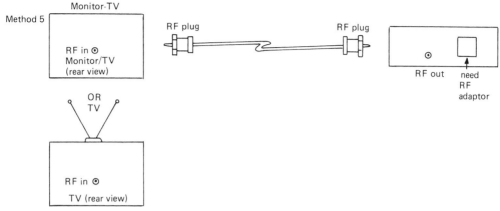

Figure 25-7 *(continued)*

VI. Notes on editing

Postproduction editing is a long, tedious process, even with sophisticated equipment. Therefore, before doing a production, choose your shots and angles carefully. Electronic editing is uncommon on open-reel tape. Mechanical editing is most unsatisfactory. If you must use an open-reel recorder, you will need to transfer the information to a cassette format. You can then edit by using two VCRs, one for playback of the old tape and one for recording the edited version, and at least one monitor or TV to view your picture. At least one of the VCR units should have automatic editing capability to help obtain invisible joining of two segments.

The units are most often connected as follows:

Playback machine (VCR 1)	VIDEO OUT/ AUDIO OUT
to	to
Recording machine (VCR 2)	VIDEO IN/ AUDIO IN

To edit, view one or more tapes on VCR 1, select the desired segments, decide what order they should be in and write out your sequence or editing plan. Keep track of the location of your segments by using the VCR tape counter and use a stopwatch to time the segments. Then play VCR 1 and record the new tape on VCR 2. Backtrack VCR 1 to slightly before each portion to be placed on the new tape so it will be recorded smoothly onto VCR 2.

Usually picture and sound are edited simultaneously. On most VCRs you can also dub sound later.

Tapes have several tracks of information. In addition to the video track, there may be one or two audio tracks, depending on machine capability, and a control track. The control track controls the speed of the VCR and the rate at which the tape is moved through the unit.

You may dub new sound onto a tape by plugging a microphone or other sound input into the recorder. Where the machine only provides one audio channel, when new sound is added the old sound is erased and replaced by the dubbed sound. With two-channel machines, you may add to the existing sound. This is useful for bilingual programs, or for adding music or commentary to an existing sound track. If you think you will be adding sound later, where you have two-channel capability record the original sound on Channel 2 of your recorder and the later sound on Channel 1. When you play back the sound, use the MIX switch.

There are basically four kinds of edits:

1. *Live assemble edit.* You correct a mistake in the studio at the time of performance. Coordinate the tape-recorder operator, camera people, and talent.
2. *Insert edit.* Here, a new video portion is inserted into a completed program. In this case, the camera is set up, the performer is cued, and the insert is timed to the recorded audio. A prerecorded video segment may also be inserted.
3. *Audio-only insert.* You can have a new audio timed to match the video.
4. *Segment assemble editing.* Segments of a recorded presentation are rearranged to make up a program with proper sequencing.

VII. **Erasing tape**

A. *To erase the entire tape:*
 1. Use a bulk eraser.
 2. Remove the camera or monitor TV connections and run the recorder in the RECORD mode. This will produce some static on the tape.
 3. Where available, turn the audio and video controls on the recorder to MANUAL, then down to the "0" level. Press RECORD. This will create a blank picture with no static.

B. *To erase tape partially:*
 Disconnect all incoming signals and press RECORD on the recorder for the portion you want to erase.

VIII. **Compatibility of equipment and tape formats**

Camera and recorder. A camera from one manufacturer is not always useable with a recorder of another manufacturer. Check with your dealer.

Tape formats. When the tape formats are not physically compatible or interchangeable, you can transfer formats. Two recorders of different formats can be connected with appropriate cables to make the transfer.

A. *¾-inch cassette format:*
 The equipment and cassette tapes for the ¾-inch cassette format are fully compatible and interchangeable. However, ¾-inch tapes are not interchangeable with ½-inch tapes.

B. *½-inch open-reel tapes:*
 ½-inch open-reel tapes can be used on any ½-inch open-reel machine.

C. ½-inch cassette format equipment and tape:

The ½-inch VHS and Beta cassette formats are not interchangeable with ½-inch open-reel. The VHS and Beta formats are also not compatible with each other, or with other less popular cassette formats. Within each format, make sure the playback unit is compatible for the speed at which your tape was recorded.

1. Video Home System (VHS) is not interchangeable with the Beta system, as mentioned above. Some VHS manufacturers are Panasonic, JVC, MGA, Sylvania, Hitachi, Magnavox, Quasar, RCA, GE, Zenith and Sharp. Most machines can record and play back in two-, four-, six-, and eight-hour modes.

    Recording in one- and two-hour modes provides better quality than recording at longer times.

2. The Beta system is not interchangeable with the VHS system. Some Beta brands are Sony, Sanyo, Toshiba, and Sears. Zenith no longer carries Beta. Compatibility and interchangeability is as follows, within Beta:

    • Beta 1—one-hour only recording and playback; standard Beta format for industrial use.

    • Beta 2—two-hour recording and playback; standard Beta format for home use.

    • Beta 3—three- to five-hour recording and playback, according to the length of tape used.

    More recent home Beta systems:

    • Two- to five-hour recording

    • One-, two-, and three-hour playback

    • Interchangeable with Beta 1, Beta 2, Beta 3

GLOSSARY

Antenna Device for transmitting and/or receiving electromagnetic waves (radio and/or video signals) through space. A *directional antenna* is one that receives from only one direction and helps to eliminate ghosts. A *main antenna* is an individual antenna in a single hookup of a private house or an individual roof antenna in a school. A *central antenna* is a master antenna for a whole building with outlets in each room.

Automatic gain control An electronic circuit that controls input levels or intensities so that the outputs stay at predetermined levels.

Backlight Light placed above and behind the subject. It is used to visually separate the subject from the background.

Broadcasting Open-circuit (rather than closed-circuit) transmission of video and/or audio signals over the air. Regulated by the Federal Communications Commission.

Built-in tuner Special feature on cassette recorder that works like a TV set without a picture tube (see Tuner).

Burned-in image Unwanted image that persists on the camera after the camera has been turned to a different scene.

CCD *See* Charged couple device.

CCU See Camera control unit.

Cable TV (CATV) System used to pick up network programs and transmit them via cable to subscribers in remote locations. These homes are linked to a common cable system unlike ordinary reception, which is received by privately owned antenna.

Camcorder All-in-one camera and recorder unit.

Camera Control Unit (CCU) Separate box, sometimes connected to the camera, which controls the camera's electronic functions. When controls are separate, the camera itself can be more compact.

Capstan servo system Electronic circuit that precisely controls the speed of the videotape. It gets its information from two sources: the sync (or control) track to ensure stable playback, and the incoming video signal to ensure proper recording.

Cartridge Videotape format that consists of one enclosed reel of tape. The take-up reel is in the machine. Therefore, the tape must be completely rewound before being removed from the machine. Available in ½-inch width but no longer in general use.

Cassette Videotape format that consists of two enclosed reels. Available in ¾-, ½-, and ¼-inch widths, to be used with machines handling those respective formats.

Charged Couple Device (CCD) Small solid-state chip useable for video camera pickup. The CCD makes a camera more compact and economical.

Clogged head Video recorder head that has a buildup of oxide. This causes improper recording and playback (noise breakup, loss of picture) and possible damage to the videotape.

Closed-circuit Television (CCTV) Video transmission via cable rather than air. The two types of CCTV are:

1. *One-room setup.* The viewer can start, stop, or replay programs on fairly portable equipment.

2. *Central location for recording and playback, with programs transmitted via cable to many rooms.* The user usually has no control over start, stop, and replay of programs.

Coaxial cable Cable that conducts RF (radio frequency) signals. The cable connects a monitor, monitor TV, or TV set to a cassette recorder. It also connects TV stations for network or CATV transmissions.

Cue Signal indicating the start of action by a performer.

Cut Order by a director to stop production, or technique requiring an abrupt change of scenes. The opposite techniques are fade and lap dissolve.

Degaus To erase or demagnetize videotape.

Dew indicator Indicator on the recorder. When activated, it shows that unit will not function because it has moisture. Moisture occurs particularly during sudden temperature changes, such as when the equipment is moved from a very cold to a warm environment. Allow the machine to warm up until the dew indicator disappears. Leave machine on without cassette.

Dissolve See Fade, Lap dissolve.

Drop out Areas on the videotape that have lost the picture because of deterioration or damage to the tape surface. Visible on the screen by a white or black streak.

Dubbing The process of adding a separate or different sound track to a video recording. With audio dubbing, the picture is not erased. Also, the process of making a tape copy.

Educational television (ETV) Describes programs of a general educational nature.

Normally not designed for direct classroom use. See Instructional television (ITV).

Electronic editing Editing of video without physically cutting tape.

Electronic Industries Association of Japan (EIAJ) Organization which has set up a number of standards for manufacturers to standardize equipment and provide interchangeability. Applies to ½-inch open-reel equipment.

External sync generator This unit provides a pulse that keeps two or more cameras electronically linked.

Fade Gradual change in picture. A *fade-out* occurs when the picture fades to black, such as at the end of a program. From black to picture is a *fade-in*.

Fill light A complement to keylight, but more diffused, fill light is used to reduce shadows. It is placed near the camera on the opposite side of the keylight, and at the same height as the camera.

Film chain System whereby there are at least three separate film images to be used one at a time for televising by a single camera. The chain may include 16mm and super 8 film, as well as 35mm slides (see Multiplexer).

First generation Tape as originally recorded, not a copy.

Ghosts Light, second image of the picture caused by poor reception.

Head Device used in video and audio recorder-players that scans tape for recording and playback information.

Helical scan Process of video magnetic tape recording. Equipment is affordable and operable by the average business, educational institution, and individual consumer. Uses ¾-inch, ½-inch, and ¼-inch videotape formats.

Idiot cards Also called cue cards. These cards contain the script so the performer can recall lines while performing.

Instructional television (ITV) Use of TV for educational purposes, particularly for classroom courses or enrichment.

Keylight Principal light on the main subject. It is often a spotlight, might be higher than camera level, and is usually placed to the left or right of the camera.

Kinescope Special 16mm film of a TV program made from a monitor or TV screen.

Lap dissolve When one image is faded over another, holding for a period before fading out. This is used to create mood. Also called dissolve.

Monitor Equipment to view video recording and playback on a picture tube. Looks like a TV set but cannot receive broadcast channels. For use in closed-circuit television (see Monitor TV).

Monitor TV set (monitor–receiver) Equipment that has both monitor and television capabilities. Monitor: viewing of video recording and playback of closed-circuit and off-the-air broadcasts. TV: viewing of broadcast channels.

Montage Composite picture made by combining several separate pictures, also the production of a rapid succession of images to illustrate an association of ideas.

Multiplexer Device used in a film chain. Coordinates film and slides when used for a video camera.

Noise Picture or sound interference that ruins the quality of the recorded action. Source of difficulty may be in the equipment or from an external cause.

Oxide Magnetic particles on the tape.

Plugs Connectors at the end of a cable.

Power Pak Box (transformer, AC adaptor) Device that decreases AC voltage to a low DC voltage. The power pak box allows battery-operated equipment to be used on AC power.

Quadruplex Process of video magnetic tape recording that has been used by television stations and other professionals. The equipment is large, heavy, expensive, and uses 2-inch-wide videotape.

RF Adaptor Unit in a video recorder to allow playback of taped programs on a regular TV set. The unit, whether built-in or bought separately, converts the recorded video and audio signals into a low-power TV broadcast frequency.

RF Converter See RF adaptor.

RF Modulator See RF adaptor.

Receiver Another term for TV set.

Signal Information, whether audio or video, converted into electrical impulses.

Skew control Recorder control that affects tape tension. Using this control on playback may straighten a bend at the top of the picture.

Slate Thirty-second videotape segment identifying a program about to be recorded. Included can be date, title, actors, and take. A reference point on the tape used for playback or editing.

Snow White dots covering the picture or blank screen, indicating poor transmission.

Streaking Abnormal picture condition in which objects are elongated horizontally.

Super Superimposition of scenes. Think of a speaker advertising a product, with the phone number added to the screen.

Sync (Synchronization) Keeping video and audio tracks in proper timing with each other.

Take Shooting of a scene. A scene could be reshot several times, resulting in several takes.

Tape leader Part of the videotape that comes before and/or after a program. It is used for winding the tape around the spools and for providing lead time before a program.

Tearing Abnormal picture condition in which horizontal lines move irregularly.

Tracking control Recorder control used to align heads with tape path. Useful when a cassette tape is played back on a machine different from that on which it was recorded. Incorrect tracking can show up on playback as tear or sideways pull of picture. Tracking control must be at center position when recording.

Transformer See Power pak box.

Tuner Device that brings in the broadcast TV signal. It does not contain a picture tube. When built into or added to a video recorder, a TV broadcast can be taped without connecting the recorder to a monitor TV set.

UHF Ultra high frequency. Also a type of connector.

VCR Video cassette recorder. Uses videotape enclosed in a plastic container or housing.

VTR Videotape recorder. Uses open-reel format, but the term sometimes is used to apply to cassette format as well.

Videotape Tape on which both sound and picture are recorded at the same time.

Vidicon tube One type of picture pickup tube used in the nonprofessional TV camera that converts visual information to electronic information for the recorder. Keep away from direct sunlight. Modified updated versions of the vidicon are the Plumbicon and Saticon tubes. Shortly, however, camera tubes will be replaced by solid-state chips (CCDs) which are more rugged.

Bibliography

Bensinger, Charles. *The Video Guide.* New York: Charles Scribner's Sons, 1981. Covers many types of video formats, explaining technically and pictorially how each operates.

Birnbaum, Hubert C. *Amphoto Guide to Cameras.* Garden City: American Photographic Publishing Co., Inc., 1978. Covers most types of camera formats and their features in summary form.

Carraher, Ron. *Electronic Flash Photography.* New York: Van Nostrand Reinhold, 1980. Gives a thorough explanation of how electronic flash works. The book is well illustrated and is useful to both beginners and advanced flash users.

Dolan, Edward F. Jr. *It Sounds Like Fun: How to Use and Enjoy Your Tape Recorder and Stereo.* New York: Julian Messner, 1981. A simple explanation of the workings of tape recorders, and equipment with which they are used.

Eboch, Sidney C., and George W. Cochern. *Operating Audio-Visual Equipment* (2nd ed.). New York: Harper and Row, 1968. Provides clear, schematic diagrams for operation of tape recorders, record players, and still-projectors and moving-picture projectors.

Englebardt, Stanley L. *Miracle Chip: The Microelectronic Revolution.* New York: Lathrop, Lee and Shepard Books, 1979. Explains how we came to the technology of miracle chips, where they are used, and what the future may look like with the use of these chips.

Fuller, Barry, Steve Kanaba, and Janyce Brisch-Kanaba. *Single-Camera Video Production.* Englewood Cliffs, N.J.: Prentice-Hall, Inc., 1982. Explores in readable detail how to produce a video program, including making sets, designing graphics, and setting up lighting.

Griffin-Beale, Christopher, and Robyn Gee. *TV and Video.* London: Usborne Publishing, Ltd. A slim volume, almost totally composed of colorful drawings, that explains very simply the processes of TV and video. Very cleverly put together!

Grimm, Tom, and Michelle Grimm. *The Good Guide for Bad Photographers.* New York: New American Library, 1982. Good information on photographic composition in addition to basic camera and flash usage.

Hawkins, John, and Susan Meredith. *Audio and Radio.* Companion volume to Griffin-Beale, *TV and Video* (see above).

Heinich, Robert, Michael Molenda, and James Russell. *Instructional Media and the New Technologies of Instruction.* New York: John Wiley and Sons, 1982. Uses a textbook approach to the selection and use of equipment and operating techniques. Includes theoretical justifications. Useful for teachers. Covers a wide range of equipment.

Kemp, Jerrold. *Planning and Producing Audio-Visual Materials* (4th ed.). New York: Harper and Row, 1980. Relates to steps in actually producing audio-visual programs. Gives useful detail.

Kettelkamp, Larry. *Lasers: The Miracle Light.* New York: William Morrow and Company, 1979. Gives basic information on how lasers work and how they are used.

Kodak. Eastman Kodak publishes easy-to-read paperback books on a variety of photographic techniques. As an example, "Movies with a Purpose: A Teacher's Guide to Planning and Producing Super-8 Movies for Classroom Use." Check with your local camera dealer.

Kybett, Harry, and Peter L. Dexnis. *Complete Handbook of Home Video Systems.* Reston, Virginia: Reston Publishing Co. (a division of Prentice-Hall, Inc.), 1982. Technical, sophisticated explanations of the workings of ½-inch VCRs.

Langford, Michael J. *The Camera Book.* New York: Ziff-Davis Publishing, 1980. Detailed information on the use of 35mm cameras, metering, lenses, lighting, and so on. Clear, concise, and well illustrated with large, colorful drawings.

Mercer, John. *An Introduction to Cinematography* (2nd ed.). Champaign, Ill.: Stipes Publishing Co., 1979. A clear, simple, basic textbook on making movies. Includes both super 8 and 16mm information.

Muse, Ken. *Photo One, Basic Photo Text.* Englewood Cliffs, N.J.: Prentice-Hall, Inc., 1973. Outstanding basic photography information in cartoon format. Equips the beginner well, while keeping you entertained.

Pinkard, Bruce. *The Photographer's Bible.* New York: Arco Publishing, Inc., 1983. A handy, hardcover reference written in alphabetic, encyclopedic format.

Renmore, C.D. *Silicon Chips and You.* New York: Beaufort Books, Inc., 1980. Basic development of chip technology to the present, along with projected applications.

Roman, James W. *Cablemania: The Cable Television Sourcebook.* Englewood Cliffs, N.J.: Prentice-Hall, Inc., 1983. Covers concepts, procedures, programming, regulation and distribution of cable, along with problems.

Schroeder, Don, and Gary Lare. *Audiovisual Equipment and Materials: A Basic Repair and Maintenance Manual.* Metuchen, N.J.: Scarecrow Press, 1979. In addition to basic information on procedures and some problem-solving for equipment, the book is useful for learning about minor electrical repairs.

Sussman, Aaron. *The Amateur Photographer's Handbook* (8th ed.). New York: Thomas Y. Crowell Company, 1973. Provides much detail in clear language relating to still photography. It is primarily for the serious photographer, and does not deal much with instant-load or instant-print photographic formats.

Wordsworth, Christopher. *The Movie Maker's Handbook.* New York: Ziff-Davis Publishing, 1979. Similar to Langford, *The Camera Book* in format (see above). Detailed with excellent drawings and photographs.

Wyman, Raymond. *Mediaware* (2nd ed.). Dubuque, Iowa: Wm. C. Brown Co., 1976. Gives theoretical background as well as technical information on many types of equipment. Good chapters on electricity, sound, magnetism, and optics. Textbook approach useable by beginning audio visual technicians.

Zelmer, A.C. *Community Media Handbook* (2nd ed.). Metuchen, N.J.: Scarecrow Press, 1979. Gives individuals and organizations ideas on how to reach the public, and how to budget and plan a presentation.

Index

AC adaptor:
  for cassette recorders, 177
  for video systems, 252
Acetate:
  definition of, 18
  writing on, 13, 15, 17–18
AFT (automatic fine tuner), 217
ALC (automatic sound level control), 176, 184, 192, 195
Alcohol, 2
Ampere, definition of, 7
Amplifiers:
  in PA systems, 209, 213
  phonograph, 186
Animation, 133
Antennas, TV, video recording and, 221, 241–43, 249
Aperture, 46, 72–74
  definition of, 74
  in filmstrip projectors, 122
  of 16mm projectors, 155, 164
  See also Exposure
ARL (automatic recording level), 176, 184, 192, 195
ASA rating, 46, 48–49, 74, 75, 133
  general guidelines for selection of, 71
Audio-only insert, 248
Autofocus, 47, 48
Autoload projectors, 147
  threading, 157–61
  See also 16mm movie projectors
Automatic film advance, 47
Automatic fine tuner (AFT), 217
Automatic gain control, 249
Automatic recording level (ARL), 176, 184, 192, 195
Automatic sound level control (ALC), 176, 184, 192, 195
Autoreverse on tape recorders, 175, 184
Auxiliary jacks, definition of, 213
Available (existing) light, 74

Backlight, 222, 249
Batteries:
  camera, 47
  for condenser microphones, 191, 193, 194
  extra, 2
  for flash, 49, 58
  general tips on 3–4
  for movie cameras, 125, 126, 133
  rechargeable, 4
  shelf life of, 3, 8
  for tape recorders, 169, 177
  in video systems, 222
  for visualmakers, 82
BCPS (beam candlepower seconds), 69–70, 74
Bell & Howell Autoload projectors, 147, 157–61
Bell & Howell 1575A projector, 156
Bell & Howell 750 filmstrip projector, 121

Beta video cassette format, 219, 221, 249
Bias, 186
Biderectional microphones, 195
Black slides, 101, 112
Brightness knob, TV, 217
Broadcasting, definition of, 249
Bulk loader for slides, 96
Burned-in image, 249

Cable television (CATV), 215, 250
  video systems and, 221–22
Camcorder, 219, 250
Camera control unit (CCU), 250
Cameras, 45–77
  cleaning, 57
  instant-loading, 45, 46
    basic elements (figure), 60
    operation, 57–59
    with visualmakers, 78–83
  operating tips for, 47–49
  problems with, 49–56
  35mm, 45
    adjustable vs. automatic, 46–47, 59–63
    operation, 59–74
  See also Film; Flash; Movie cameras; Video system—cameras in
Capstan roller, 184
Capstan servo system, 250
Cardioid microphones, 195
Carrier of thermal transparency maker, 41, 44
Cartridges:
  phonograph, 187–90
  video, 219, 250
Cassettes, *see* Tape recorders—cassette; Video systems—videotape in
Charged couple device (CCD), 250, 252
Chartex, 26
Circuit, definition of, 7
Closed-circuit television (CCTV), 250
Coaxial cable, 250
Color balance, 74
Color film, 71–72
Color knob, TV, 217
Color level control, preset, 217
Compressed air, 2, 74–75, 133
Condenser lenses, 99
  in filmstrip projectors, 122
  of 16mm projectors, 164
Condenser microphones, 191, 193–95
Connectors, *see* Patchcords; Video systems—connections in
Contact prints, 75
Contrast control, TV, 217
Copy stands, 84–85
  for visualmakers, 78, 81–83
Copyright laws, 1, 218
Cords, electrical, 8

257

Core of laminating film, 40
Couplers, magnetic, 198
Crystal microphones, 195
Cue control, tape recorder, 184
Cues in video production, 250, 251
Cut:
   in movie editing, 133
   in video, 250

Darkness dial of thermal transparency maker, 42, 43
Degaussing, 222, 250
Depth of field, 75
Dew indicator, 250
Digital recording, 190
Digital tone, 112
Digital TV, 214, 217
DIN designation, 75, 207
Dissolve units, 96, 99–112
   basic connections with projectors (figure), 110
   operating procedures for, 107–12
   problems with, 102–7
Dissolves, video, 250, 251
Ditto machine, spirit masters for, 41
Dolby system, 184
DRIVE laminating control, 35, 40
Drop out, videotape, 250
Dry-mount press, 22–26
Dry-mounting tissue, 26
Dubbing, video, 248, 250
Dynamic microphones, 195

Editing:
   movie, 125, 133
   video, 247–48
Educational television (ETV),250–51
8mm video format, 219
Eiki 16mm projectors, 156, 162
Ektachrome film, 83
Ektagraphic projector, 99
Electric eye, 75
Electricity, general tips on, 6–8
Electronic Industries Association of Japan (EIAJ), 251
Emulsion:
   on camera film, 75
   on movie film, 133, 164
Equipment:
   accessory list of, 1–2
   cleaning and servicing of, 9
   handling of, general tips on, 2–3
   model numbers of, 9
   repair of, 9
   storage of, 8
   See also specific equipment
Erasure, tape:
   audio, 177
   video, 222, 248
Establishing shot, 133
Exacto knives, 2
Exciter lamp, 149, 150, 155, 164
Existing (available) light, 74
Exposed film, 75
Exposure:
   automatic setting of, 46–48
   changing, 48–49, 72–74
   definition of, 75
   over-, 76
   under-, 77
Exposure meters, *see* Light meters
Extender arm, camera, 76
External sync generator, 251

f stop, 46, 72–74, 76
   on video cameras, 220, 223
Fades:
   movie cameras and, 133
   video, 251
Fast motion, 133
Federal Communications Commission, 249

Feedback:
   definition of, 213
   with microphones, 193–94
   with PA systems, 210
Fill light, 222, 251
Film, laminating, 36–40
Film, movie, 125
   8mm, 138–39
   problems with, 127, 128, 131
   running time of, 132
   selection of, 131
   in 16mm projection, 145, 151–54
   in super 8 projection, 138–39, 141–43
Film, still-camera, 2
   care of, 48
   instant-loading, 58–60
   print vs. slide, 59
   problems with, 49, 50, 53, 55, 56
   shelf life of, 8
   35 mm:
      color, 70–72
      loading and unloading, 63–65
      selection of, 70–71
   for visualmakers, 83
Film chain, 251
Film cutter/trimmer, 164
Film gate, 145, 164
   in filmstrip projectors, 123
Film indicator, 125, 131, 133
Film loop projectors, 135–37
Film path/channel, definition of, 164
Filmstrip adaptor for slide projector, 96
Filmstrip projectors, 113–23
   filmstrip–sound synchronization problems with, 116–17
   operating procedures for, 117–22
   parts of (figure), 118
   phonograph combination with, 190
   sound problems with, 117
   types of, 113–14
   visual problems with, 114–16
Filters:
   movie-camera, 131
   still-camera, 71–72
First generation video recording, 251
Flags, 164
Flash, 48, 49
   dedicated, 67–68, 75
   electronic, definition of, 75
   with instant-loading cameras, 58–59
   problems with, 51–53, 55, 56
   with 35mm cameras, 65–70
   with visualmakers, 79–82
Flood lamps, 71
Focal length, 76
Focus:
   camera, 46–48, 54
   with dissolve units, 105
   of filmstrip projectors, 116, 119–21
   of 16mm projectors, 148
   of slide projectors, 87, 90, 94
   of super 8 projectors, 140
   video, 223
Fotoflat tissue, 26
fps (frames per second), 129, 134
Framer of 16mm projector, 155, 164
Freeze action, 48, 73–74
Fresnel lens, 15, 18

Gate index, 99
General Electric, 12, 249
Ghosts, 216, 251
Graflex 900 projector, 156–57
GTE Products Corporation, 12

Heads:
   tape-recorder, 167, 169, 178, 183
   video, 234, 250, 251
Headsets, 197–200

Index   259

HEAT control in lamination, 35, 40
Heat filters, 2
    for filmstrip projectors, 123
    for slide projectors, 89, 99
Heat lifting, 22, 24, 26
Heating shoes, 40
Helical-scan video, 218, 251
Hitachi video, 249
Hot shoe flash connection, 66, 67, 76
Hue (of TV picture), 217

Idiot cards, 251
Idler roller, 184
Impedance of microphones, 191, 195
Index counter, 184
Input jacks, definition of, 213
Insert edit, 248
Instructional television (ITV), 251
ISO rating, 46, 48–49, 76, 133
    general guidelines for selection of, 71
    movie film and, 134

Jackboxes, 197–201
Jacks:
    adaptors for, 200
    definition of, 201, 202, 213
    for headsets, 197, 198
    microphone, 191, 203
    PA-system, 211–12
    in patching procedures, 202–7
    phonograph, 186, 189
    tape-recorder, 184
    types of, 207
    video, 237
        See also Video systems—connections in JVC video, 249

Kelvin scale, 76
Keylight, 222, 251
Kinescope, 251
Kodachrome film, 59, 60, 83, 86
Kodacolor film, 59, 60, 83
Kodak Carousel slide projector, 93, 99
Kodak Instamatic camera, visualmaker kits for, 78–83
Kodak Pageant 16mm projector, 151, 162

Lamination, 27–40
    definition of, 26, 27
    with dry-mount press, 22–24, 26
    handling of film in, 36–40
    machine cleaning in, 40
    machine features in, 27, 34
    machine operations in, 35–36
    machine problems in, 32–33
    materials unsuitable for, 27
    product problems in, 28–32
    with 3M Thermofax, 41
Lamps, 1, 10–12
    codes of, 12
    flood, 71
    shelf life of, 8
    of 16mm projectors, 144, 145, 154
        See also Lighting
Lap dissolve in video, 251
Leader:
    on movie film, 164
    on still film, 76
    on videotape, 252
LED (light-emitting diode), 167, 184
Lens barrel, 123
Lens cleaners, 2
Lens housing of movie projector, 164
Lens speed, definition of, 76
Lenses:
    condenser, 99
    for filmstrip projectors, 121–123
    of movie cameras, 133
    of 16mm projectors, 147–49, 163, 164
    of slide projectors, 89, 96, 99

of still cameras:
    care of, 47, 57
    SLR, 46
of video cameras, 223
    zoom, 125–26, 134
Light meters:
    automatic settings and, 47
    definition of, 77
    how to use, 61
    problems with, 50–51, 55
Lighting:
    color film and, 71–72
    for copy stands, 84–85
    flood, 71
    movie film and, 131
    tungsten, 77
    video, 222–23
        See also Flash; lamps
Linear tracking, 190
Listening stations, 197–201
Live assemble edit, 248
Loop projectors, 135–37
Loop restorer, 164
Loops in threading of 16mm film, 155, 164
Loudspeakers, PA-system, 209–13

Magnavox video, 249
Magnetic couplers, 198
Magnetic sound system, 138, 149, 150, 163, 164
Manual override, 47, 76
Manual projectors:
    threading, 161–63
        See also 16mm projectors
Master volume control, PA, 212, 213,
MGA video, 249
Microphones, 191–96
    maintenance of, 195
    operating tips on, 191–93
    in PA systems, 209–12
    problems with, 193–94
    for tape recorders, 168, 169, 176, 185
    types of, 195
Mini plugs, 207
Mixers, 192, 195
Mixing, video, 248
Mode, definition of, 184
Model numbers, 9
Monaural (mono) records, 187, 190
Monitor switch on tape recorder, 176
Monitors, see Video systems—monitor/TV in
Montage, 251
Mounts for slides, 88, 90, 105
Movie cameras, 124–34
    cleaning, 133
    filters for, 131
    operating procedures for, 129–31
    operating tips on, 124–25
    problems with, 126–29
    16mm, 124
    sound and, 125, 127, 129, 132–33
        See also Film, movie
Movie projectors, see 16mm movie projectors; Super 8 loop projectors; Super 8 movie projectors
Moving-coil microphones, 195
Multiplexer, 251

Needles, phonograph, 186–90
Newsprint, blank, 2
Nickel cadmium batteries, 49
Nip of laminator, 40
Noise, video, 251

OFF laminator control, 35, 40
Omnidirectional microphones, 195
Opaque projectors, 19–21
Optical sound system, 138, 149, 150, 163, 164
Outlets, wall, 6–7

Output jacks, definition of, 213
Overexposure, 76
Overhead projectors, 13–18
Oxide, 250, 251

Pan shots, 129, 134
Panasonic video, 249
Parallax, 76
Patchcords (connectors), 202–8
  definition of, 184, 202
  format transfer with, 203–7
  replacing plugs on, 207, 208
Phono jacks, 207
Phonographs, 186–90
Photo flood bulbs, 85
Pinch roller, 184
Plugs, 184
  definition of, 201, 202, 251
  headset, 198–201
  patchcord, replacement of, 207, 208
  in patching procedures, 202–7
  types of, 207
  video, 235, 236
  See also Video systems—connections in wall, 7–8
Platen:
  in dry-mount press, 25–26
  in opaque projection, 19–21
  in overhead projection, 21
Plumbicon tube, 252
Pointer in opaque projector, 21
Power Pak Box, 252
PREHEAT control in laminator, 35, 40
Presentation, planning of, 1
Programmed tape for slide advance, 96–97
Projector-screen distance, 4–5
Projector lens, 123
Projectors, *see specific types of equipment*
Public address (PA) systems, 209–13
  types of, 209
Pulldown claw, 164
Pulse, 112, 123
  on audio tape, 184–85
Pushing film, 77

Quadruplex video, 218, 252
Quasar video, 249

Rangefinder, camera, 77
RCA jacks, 207
RCA 16mm projector, 162
RCA video, 249
Record players, 186–90
Records, phonograph, 186–88
Reel-to-reel tape recorders, *see* Tape recorders—reel-to-reel
Registration slides, 112
Remote control:
  with dissolve units, 106
  of slide projector, 91, 96, 99
Remote microphones for tape recorders, 168, 169, 176, 185
Repair of equipment, 9
Reverse (projector function), 164–65
Rewind (projector function), 165
RF adaptor/converter/modulator, 220, 252
Rollers, laminator, 34
Rubber roller, 185
Rulers, 2
RUN laminating control, 35, 40

Sanyo video, 249
Saticon tube, 252
Screens:
  distance from projector, 4–5
  general tips on, 5–6
Screwdrivers, miniature, 2
Sears video, 249

Segment assemble editing, 248
Select button/bar of slide projector, 99
Servicing of equipment, 9
Shaft, laminator, 40
Sharp video, 249
Shelf life, 8
Shoes, laminator, 40
Shutter speed, 46, 48, 72–74
  definition of, 77
  See also Exposure
Side plates, laminator, 40
Signal, definition of, 252
Singer 16mm projectors, 156
Single lens reflex (SLR), 46, 77
16mm movie cameras, 124
16mm movie projectors, 144–65
  cleaning, 163
  film problems with, 151–54
  light and lamps of, 145, 146
  operating problems with, 145–46
  parts of, 144, 155–56
  picture problems with, 146–49
  rewinding, 163
  running time and reel size for, 163
  sound systems of, 147, 149–51, 155, 163, 164
  threading, 156–63
Skew control, 252
Slate, 252
Slide projectors, 86–99
  cleaning, 99
  continuous projection with, 98
  features of, 87–88
  lenses for, 89, 96, 99
  operating procedures for, 93–98
    special slide-advance devices, 96–98
  power problems with, 88
  slide and projector problems with, 89–92
  sound problems with, 93
  sound synchronization with, 87, 92
  types of, 87–88
  See also Dissolve units
Slides:
  formats of, 86
  mounts of, 88, 90, 105
  See also Transparencies
Slot-load projectors:
  See also 16mm movie projectors
Slow motion, 129, 134
Snow, 252
Sony video, 218, 249
Sound drum, 145, 155, 165
Sound systems:
  film, 138, 141, 147, 149, 150, 163, 164
  phonograph, 186–90
  See also Microphones
Speakers:
  in PA systems, 209–13
  phonograph, 186, 190
Spirit masters for ditto machine, 41
Splicing, 134, 139, 145
Sprocket guard, 165
Sprocket holes:
  film, 77
  filmstrip, 123
Sprocket wheel of filmstrip projector, 123
Sprockets, camera, 77
Stack loader for slides, 96
Stage:
  in opaque projection, 21
  in overhead projection, 18
Stage glass, 18
Stereo headsets, 198
Stereo recording, 185
  phonograph, 187, 190
Storyboard, 125, 134
Streaking, 252

Index **261**

Strobe, 77
Stroboscopes, 190
Submini plugs, 207
Super, 252
Super 8 loop projectors, 135–37
Super 8 movie cameras, *see* Movie cameras
Super 8 movie projectors, 138–43
   cleaning, 143
   operating problems with, 139–43
   types of, 138
Sylvania *Photographic Lamp and Equipment Guide*, 12
Sylvania video, 249
Sync cord flash connection, 66–67
Synchronization (sync):
   sound, with slide projectors, 87, 92
   of tape-recorder combinations, 174, 179–80, 185
   video, 252

Tacking iron, 23–26, 40
Take, 252
Take-up reel, 145, 165
Take-up sprocket, 165
Tape, magnetic:
   audio, *see* Tape recorders
   video, *see* Video systems—videotape in
Tape leader, 252
Tape recorders, 166–85
   cassette, 166, 175–80
     batteries for, 177
     operating procedures, 175–177
     tapes, 178–80
   with dissolve units, 107
   duplicating tapes for, 180
   erasing procedure for, 177
   with filmstrip projectors, 117, 121–22, 179
   head care, 167, 169, 178, 183
   microphones of, 168, 169, 176, 185
   operating tips for, 167–68
   playback sound problems with, 169–72
   power problems with, 169
   recording problems with, 172
   reel-to-reel, 167, 180–83
     machine speed, 182, 183
     tapes, 183
   for slide projectors, 93, 96, 174, 179
   synchronization with other machines, 174, 179–80, 185
   tape considerations, 167–68, 172–73
Tearing, 252
Television, 214–17
   cable, 215, 221–22, 250
   closed-circuit, 250
   digital, 214, 217
   problems with, 215–16
   tuning, 216–17
   with video systems, 220
   See also Video systems—monitor/TV in
Telex 16mm projectors, 156
Thermal transparency film, 41–43
   shelf life of, 8
   transferring print or image onto, 13, 15
Thermal transparency makers, 41–44
Threading knob, laminator, 39, 40
Threading lamp, 165
3M Thermofax transparency maker, 41–44
Through-the-lens (TTL) viewing, 46, 75, 77
Timer of slide projector, 96
Timer video recording, 235, 242–43
Tint (of TV picture), 217
Top glass, 18
Toshiba video, 249
Track, tape, 185
Tracking control, 252
Transformer, *see* AC adaptor
Transparencies:
   heat-lifting process for, 22, 24, 26

   overhead projection of, 13–18
   See also Slides; Thermal transparency film
Trays, slide-projector, 88, 89, 91, 93–96
   with dissolve units, 102–4
Trucking shot, 134
TTL (through-the-lens) viewing, 46, 75, 77
Tuner, built-in, 221, 223, 235, 237, 241–43, 249, 252
Tungsten film, 71, 72, 77
Tungsten lighting, 77
Tweeter, 190
Tweezers, 2

UHF (Ultra high frequency), 252
"U-matic" video cassette format, 219
Underexposure, 77
Unidirectional microphones, 193, 195–96

Verichrome Pan film, 59, 60, 83
VHS (Video Home System) format, 219, 221, 249
Video cassette recorder (VCR):
   definition of, 252
   See also Video systems—recorder/players in
Video systems, 218–52
   and cable TV, 221–22
   cameras in, 218, 220–21, 235, 248
     basic parts (figure), 238
   cleaning, 234–35
   connections in, 218, 222, 235, 236
     for playback, 245–47
     for recording, 238–39, 241–44
   editing and, 247–48
   equipment-tape compatibility in, 248–49
   erasing tape, 248
   monitor/TV in, 218, 220, 221, 234–35
     definition of, 251
   playback procedures, 244–47
   problems with, 225–33
     general, 225–27
     picture, 228–31
     recording, 227–28
     sound, 231–32
     tape, 232–33
   recorder/players in, 218–21, 234, 248
     parts of (figure), 237
   recording procedures, 222–24, 235–44
     live, 235, 238–41
     off-the-air, 235, 241–44
   types of, 218–20
   videotape in, 218–20, 222, 224–25, 233–34
     cartridge format, 219, 250
     cassette formats, 219, 220, 224–25, 248–50
     open-reel format, 219, 220, 224–26, 248

Videotape recorder (VTR):
   definition of, 252
   See also Video systems—recorder/players in
Vidicon tube, 252
Viewfinder, definition of, 77
Visualmakers, 78–83
Voltage, definition of, 7
VU meter, 167, 184, 185

Wall outlets, 6–7
Wall plugs, 7–8
Wattage, definition of, 7
White balance switch, 220
Windsocks, 125, 134
   microphone, 196
Wipe (movie technique), 134
Woofers, 190

Zenith video, 249
Zoom effect:
   movie, 129–30, 134
   video, 223